Micromechanics of Fracture and Damage

Micromechanics Set

coordinated by
Djimédo Kondo

Volume 1

Micromechanics of Fracture and Damage

Luc Dormieux
Djimédo Kondo

iSTE

WILEY

First published 2016 in Great Britain and the United States by ISTE Ltd and John Wiley & Sons, Inc.

ISTE Ltd
27-37 St George's Road
London SW19 4EU
UK

www.iste.co.uk

John Wiley & Sons, Inc.
111 River Street
Hoboken, NJ 07030
USA

www.wiley.com

Library of Congress Control Number: 2016933398

British Library Cataloguing-in-Publication Data
A CIP record for this book is available from the British Library
ISBN 978-1-84821-863-5

Contents

Notations

$\underline{z}$:	position vector
z_i:	ith component of $\underline{z}$ in a cartesian orthonormal frame
$z = z_1 + iz_2$:	complex number associated with the position vector
$U(\underline{z})$:	Airy's stress function at point $\underline{z}$
$\phi(z)$, $\chi(z)$, $\psi(z) = \chi'(z)$:	complex potentials related to $U(\underline{z})$ by [1.17]
$\phi(\underline{z})$, $\psi(\underline{z})$:	Papkovitch–Neuber (real) potentials
$z = \omega(\zeta)$:	conformal map
$\boldsymbol{\sigma}(\underline{z})$:	stress tensor at point $\underline{z}$
$\boldsymbol{\varepsilon}(\underline{z})$:	strain tensor at point $\underline{z}$
$\underline{\xi}(\underline{z})$:	displacement vector at point $\underline{z}$
$\mathbf{G}(\underline{z})$:	second-order Green's tensor for an infinite elastic (2D or 3D) continuum
$\mathbb{C}$:	fourth-order stiffness tensor
$\mathbb{C}^s$:	solid stiffness tensor
$\mathbb{S}^s$:	solid compliance tensor
$\mathbb{C}^{\text{hom}}$:	macroscopic (homogenized or effective) stiffness tensor

k: elastic bulk modulus
μ: elastic shear modulus
ν: Poisson ratio
E: Young's modulus
k^s: elastic bulk modulus of the solid phase
μ^s: elastic shear modulus of the solid phase
ν^s: Poisson ratio of the solid phase
E^s: Young's modulus of the solid phase
$\mathbf{1}$: second-order unit tensor
$\mathbb{I}$: symmetric fourth-order unit tensor
$\mathbb{J}$: fourth-order spherical projector
$\mathbb{K}$: fourth-order deviatoric projector ($\mathbb{I} = \mathbb{J} + \mathbb{K}$)
$\mathbb{\Gamma}^{\infty}(\underline{z})$: Kernel of the Green operator (infinite space)
$\mathbb{\Gamma}(\underline{z}, \underline{z}')$: Kernel of the Green operator (finite domain)
Γ: Green operator
$\mathbf{\Sigma}$: macroscopic Cauchy stress tensor
$\mathbf{E}$: macroscopic strain tensor
$\mathbb{P}$: fourth-order Hill's tensor
J_o: zero-order Bessel function
$\mathcal{Y}(t)$: Heaviside function
$\mathcal{F}_s$: Fourier sine transform
$\mathcal{F}_c$: Fourier cosine transform
$\mathcal{H}$: Hankel transform
K_I: stress intensity factor (mode I)
K_{II}: stress intensity factor (mode II)
G: energy release rate
G_c: fracture energy
d: crack density parameter
$\mathcal{Y}_d$: driving force of damage propagation

Preface

And it shall come to pass, while my glory passeth by,
that I will put thee in a clift of the rock,
and will cover thee with my hand while I pass by.

Exodus 33:22

An examination of the literature devoted to cracked media reveals that there are two main options for the geometrical modeling of cracks:

– the first option [GRI 21] consists of the idealized representation of the crack as two parallel faces (segments in plane strain/stress conditions or plane surfaces in three dimensions [3D]). The usual approach consists of adopting stress free boundary conditions on the crack faces. The two faces asymptotically coincide in this mathematical idealization and the displacement undergoes a discontinuity across the crack line (respectively, surface). Indeed, the displacement vectors of two material points located on each face at the same geometrical point in the initial configuration can differ from one another. Clearly, the discontinuity of the displacement field is a consequence of the idealization of the crack as a geometrical entity having a measure equal to zero in the integral sense. For the same reason, the stresses at a crack tip are singular, which has led to the introduction of the well-known concept of stress intensity factors. This first model is referred to throughout the book as the Griffith crack model. It will be presented in two-dimensional conditions (plane strain/stress), as well as in 3D conditions;

– as a second option, the crack is represented as a flat cavity. For instance, it will be a flat ellipse in plane strain/stress conditions, or a flat spheroid (or ellipsoid) in 3D, characterized by an infinitesimal aspect ratio. Consequently, the mathematical measure (in the integral sense) of the crack is infinitesimal but non-zero. This point of view represents the cracked medium as a heterogeneous material and the crack itself as an inhomogeneity in the sense of the homogenization theory. This geometrical description will therefore be referred to as the inhomogeneity model. As long as the aspect ratio has a small but non-zero value, the latter model warrants the ability to define a continuous extension of the displacement field in the crack cavity, as done classically in micromechanics of porous media. It also avoids the occurrence of stress singularities.

The very existence of two geometrical models for the same physical entity raises the question of their consistency. As pointed out above, one model induces mathematical singularities while the other model preserves the continuity of the displacement field and the absence of stress singularity, provided that the aspect ratio remains infinitesimal but non-zero. This of course might erroneously suggest that the two models are not compatible. In fact, the consistency must be examined in an asymptotic sense, when the crack aspect ratio tends to 0. It will be shown that the two models yield perfectly consistent estimates in terms of effective elastic properties. A thorough comparison of the local stress, strain and displacement fields is also proposed.

The book is organized as follows:

– Chapter 1 presents some mathematical tools of the theory of linear elasticity, which will be useful in forthcoming developments. Beginning with plane elasticity, the method of the Airy function is recalled. Biharmonic stress functions can be generated in a systematic way by means of the complex potential approach, which is also briefly presented. The method of the Airy function will be implemented in the framework of each of the two geometrical models;

– in view of application to the inhomogeneity model, Chapter 2 first introduces the Green's function. This paves the way for a presentation of the so-called inclusion and inhomogeneity Eshelby problems. Indeed, the solutions of the latter requires the determination of the Hill tensor, which is defined from the derivatives of the Green's function. Eshelby's inclusion problem is a first step toward the concept of polarization. This motivates the introduction of the Green operator. These tools will be essential for the derivation of variational bounds on the effective elastic properties of microcracked media;

– Chapter 3 deals with the Griffith crack model in two-dimensional conditions. To begin with, the stress singularity at the crack tip and the stress intensity factors are introduced. Then, the complete solutions to mode I and mode II loadings are derived, based on the use of a displacement potential technique (Papkovitch–Neuber potential), which is directly presented in the context of its implementation to crack problems. This yields the corresponding stress intensity factors. Similarly, Chapter 5 deals with the Griffith crack model in 3D conditions. Again the complete 3D solutions in mode I and in shear mode are derived;

– Chapter 4 is devoted to the inhomogeneity model of crack in two-dimensional conditions. The cross-section of the crack is assumed to be a flat ellipse. Two different mathematical techniques are implemented, namely the complex potential approach of the Airy stress function and the solution to the Eshelby inhomogeneity problem. The same Eshelby-based technique is used in Chapter 6 in order to deal with 3D flat spheroidal cracks;

– Chapter 7 introduces the concept of energy release rate and presents the classical thermodynamic reasoning leading to the related criterion for crack propagation;

– the second part of the book is devoted to the effective properties of microcracked media and to damage modeling. It opens with Chapter 8, which proposes a brief introduction to the homogenization of heterogeneous elastic media. The two geometrical models for microcracks (Griffith crack and inhomogeneity model) are successively considered. These two routes are explored in Chapter 9 (Griffith crack) and in Chapters 10 and 11 (inhomogeneity model);

– Chapter 12 is devoted to the variational approach to effective properties. It first presents the energy-based definition of the effective stiffness. Then, the Hashin–Shtrikman–Willis variational approach is detailed. The discussion emphasizes the respective roles of the inhomogeneity shape (flat spheroid in the present case) and of the crack spatial distribution;

– eventually, a micromechanics-based damage constitutive law can be formulated and this is the aim of Chapter 13, which serves as a conclusion to this book. Uniqueness and stability issues concerning the damage model will be discussed.

Rock of Ages, cleft for me,
Let me hide myself in Thee.

Luc DORMIEUX
Djimédo KONDO
February 2016

PART 1

Elastic Solutions to Single Crack Problems

1

Fundamentals of Plane Elasticity

The purpose of this chapter is to present the solution to plane elasticity problems, based on the use of complex-valued potentials. An isotropic linear elastic behavior is considered (except in section 1.8).

1.1. Complex representation of Airy's biharmonic stress function

Let U be an Airy stress function, from which the stress components in plane elasticity conditions are derived according to:

$$\sigma_{xx} = \frac{\partial^2 U}{\partial y^2}; \quad \sigma_{yy} = \frac{\partial^2 U}{\partial x^2}; \quad \sigma_{xy} = -\frac{\partial^2 U}{\partial x \partial y} \qquad [1.1]$$

Let $\Pi\boldsymbol{\sigma} = \boldsymbol{\sigma} - \sigma_{zz}\underline{e}_z \otimes \underline{e}_z$ denote the projection on the plane $(\underline{e}_x, \underline{e}_y)$ of a stress tensor $\boldsymbol{\sigma}$ defined by [1.1]. It is readily proven that $\Pi\boldsymbol{\sigma}$ is given by:

$$\Pi\boldsymbol{\sigma} = (\Delta U)\left(\mathbf{1} - \underline{e}_z \otimes \underline{e}_z\right) - \boldsymbol{\nabla}(\boldsymbol{\nabla} U) \qquad [1.2]$$

This expression is useful for the derivation of the components of σ in polar coordinates as a function of the partial derivatives of U. To do so, we recall that:

$$\Delta U = U_{,rr} + \frac{1}{r}U_{,r} + \frac{1}{r^2}U_{,\theta\theta} \quad [1.3]$$

and

$$\begin{aligned}\nabla(\nabla U) &= U_{,rr}\underline{e}_r \otimes \underline{e}_r + (\frac{1}{r}U_{,r} + \frac{1}{r^2}U_{,\theta\theta})\underline{e}_\theta \otimes \underline{e}_\theta \\ &+(\frac{1}{r}U_{,r\theta} - \frac{1}{r^2}U_{,\theta})(\underline{e}_\theta \otimes \underline{e}_r + \underline{e}_r \otimes \underline{e}_\theta)\end{aligned} \quad [1.4]$$

Introducing [1.3] and [1.4] into [1.2], we obtain:

$$\sigma_{,rr} = \frac{1}{r}U_{,r} + \frac{1}{r^2}U_{,\theta\theta};\ \sigma_{,r\theta} = -\frac{1}{r}U_{,r\theta} + \frac{1}{r^2}U_{,\theta};\ \sigma_{,\theta\theta} = U_{,rr} \quad [1.5]$$

Equations [1.5] are the counterpart in polar coordinates of equations [1.1]. The compatibility condition of the strains, which reads:

$$\frac{\partial^2 \varepsilon_{yy}}{\partial x^2} + \frac{\partial^2 \varepsilon_{xx}}{\partial y^2} - 2\frac{\partial^2 \varepsilon_{xy}}{\partial x \partial y} = 0 \quad [1.6]$$

is ensured, in the case of an isotropic linear elastic behavior, by the condition

$$\Delta\Delta U = 0 \quad [1.7]$$

As a matter of fact, under plane stress or strain conditions, the assumption of linear isotropy allows to write the state equations in the form:

$$\begin{aligned} \varepsilon_{xx} &= A_{11}\sigma_{xx} + A_{12}\sigma_{yy} \\ \varepsilon_{yy} &= A_{12}\sigma_{xx} + A_{22}\sigma_{yy} \\ 2\varepsilon_{xy} &= A_{66}\sigma_{xy} \end{aligned} \quad [1.8]$$

Under plane stresses, the elastic compliances A_{ij} are:

$$A_{11} = A_{22} = \frac{1}{E}; \quad A_{12} = -\frac{\nu}{E}; \quad A_{66} = \frac{2(1+\nu)}{E} \quad [1.9]$$

Under plane strains, these relations become:

$$A_{11} = A_{22} = \frac{1-\nu^2}{E}; \quad A_{12} = -\frac{\nu(1+\nu)}{E}; \quad A_{66} = \frac{2(1+\nu)}{E} \quad [1.10]$$

In both plane strains and plane stresses, the A_{ij} satisfy:

$$2A_{12} + A_{66} = 2A_{11} = 2A_{22} \quad [1.11]$$

Combining [1.1] with [1.10] and using [1.11], we see that condition [1.6] reduces to [1.7]. Such a biharmonic function U is now considered. Let $P = \Delta U$. By definition, P is a harmonic function. Let Q denote the conjugate function, defined up to a constant by:

$$\frac{\partial P}{\partial x} = \frac{\partial Q}{\partial y}; \quad \frac{\partial P}{\partial y} = -\frac{\partial Q}{\partial x} \quad [1.12]$$

This implies that the complex-valued function $f(x+iy) = P(x,y) + iQ(x,y)$ is holomorphic, which means that the limit (with $z = x + iy$)

$$\lim_{dz \to 0} \frac{f(z+dz) - f(z)}{dz}$$

exists. Indeed, at the first order in dx and dy:

$$f(z+dz)-f(z)=\frac{\partial P}{\partial x}dx+\frac{\partial P}{\partial y}dy+i\left(\frac{\partial Q}{\partial x}dx+\frac{\partial Q}{\partial y}dy\right) \quad [1.13]$$

Using [1.12] with [1.13] yields

$$f(z+dz)-f(z)=\left(\frac{\partial P}{\partial x}+i\frac{\partial Q}{\partial x}\right)dz \quad [1.14]$$

so:

$$f'(z)=\lim_{dz\to 0}\frac{f(z+dz)-f(z)}{dz}=\frac{\partial P}{\partial x}+i\frac{\partial Q}{\partial x} \quad [1.15]$$

Following [MUS 53], consider now a primitive $\phi(z)=p+iq$ of $f(z)/4$:

$$\phi'(z)=\frac{1}{4}f(z)$$

where p and q are two conjugate harmonic functions. Therefore, we have:

$$\frac{\partial p}{\partial x}=\frac{\partial q}{\partial y}=\frac{P}{4}; \quad P=2(\phi'(z)+\overline{\phi'(z)}) \quad [1.16]$$

We can see that

$$p_1=U-px-qy$$

is harmonic, and that

$$px+qy=\mathcal{R}e(\overline{z}\phi(z))$$

Finally, let $\chi(z)$ denote the holomorphic function whose real part is p_1:

$$p_1 = \mathcal{R}e(\chi(z)) = \frac{1}{2}\left(\chi(z) + \overline{\chi(z)}\right)$$

Following these definitions, we have:

$$U = \frac{1}{2}\left(\chi + \overline{\chi} + \overline{z}\phi(z) + z\overline{\phi(z)}\right) \qquad [1.17]$$

For future purposes, let us determine the partial derivatives of U. Observing that $\partial z/\partial x = \partial \overline{z}/\partial x = 1$, we first obtain:

$$\frac{\partial U}{\partial x} = \frac{1}{2}\left(\chi'(z) + \overline{\chi'(z)} + \phi(z) + \overline{\phi(z)} + \overline{z}\phi'(z) + z\overline{\phi'(z)}\right) \qquad [1.18]$$

In turn, $\partial z/\partial y = -\partial \overline{z}/\partial y = i$ yields:

$$\frac{\partial U}{\partial y} = \frac{i}{2}\left(\chi'(z) - \overline{\chi'(z)} + \overline{\phi(z)} - \phi(z) + \overline{z}\phi'(z) - z\overline{\phi'(z)}\right) \qquad [1.19]$$

It is convenient to summarize these results in the form:

$$\frac{\partial U}{\partial x} + i\frac{\partial U}{\partial y} = \phi(z) + z\overline{\phi'(z)} + \overline{\psi(z)} \qquad [1.20]$$

with the notation $\psi(z) = \chi'(z)$.

1.2. Force acting on a curve or an element of arc

Let us consider a curve oriented by the tangent unit vector $\underline{t}$:

$$\underline{t} = \frac{dx}{ds}\underline{e}_x + \frac{dy}{ds}\underline{e}_y$$

where s denotes the curvilinear abscissa. The positive direction of the normal unit $\underline{n}$ is defined such that $(\underline{n}, \underline{t})$ is oriented like $(\underline{e}_x, \underline{e}_y)$. This being the case, we have:

$$\underline{n} = \underline{t} \wedge \underline{e}_z = \frac{dy}{ds}\underline{e}_x - \frac{dx}{ds}\underline{e}_y$$

Using [1.1], the components of the stress vector $\underline{T} = \boldsymbol{\sigma} \cdot \underline{n}$ read:

$$\begin{aligned} T_x &= \underline{e}_x \cdot \boldsymbol{\sigma} \cdot \underline{n} = \sigma_{xx} n_x + \sigma_{xy} n_y = \frac{d}{ds}\left(\frac{\partial U}{\partial y}\right) \\ T_y &= \underline{e}_y \cdot \boldsymbol{\sigma} \cdot \underline{n} = \sigma_{yx} n_x + \sigma_{yy} n_y = -\frac{d}{ds}\left(\frac{\partial U}{\partial x}\right) \end{aligned} \quad [1.21]$$

The elementary force $\underline{T}ds$ acting on ds is represented by a complex dF with real and imaginary parts $T_x ds$ and $T_y ds$. Using [1.21], this yields:

$$dF = (T_x + iT_y)ds = -i\, d\left(\frac{\partial U}{\partial x} + i\frac{\partial U}{\partial y}\right) \quad [1.22]$$

By integration, we obtain the resultant force $\underline{F}$ acting on a given arc oriented from A to B. Introducing [1.20] into [1.22], the components F_x and F_y are given by:

$$F_x + iF_y = -i[\phi(z) + z\overline{\phi'(z)} + \overline{\psi(z)}]_{sA}^{sB} \quad [1.23]$$

The boundary conditions on a loaded arc are an important application of this result. In the following, let $f(z)$ be defined as:

$$f(z) = \phi(z) + z\overline{\phi'(z)} + \overline{\psi(z)} = i\int_{AB_z} (T_x + iT_y)ds + \text{Const} \quad [1.24]$$

where the point A is fixed and z denotes the affix of point B_z. $f(z)$ is a complex representation of the resultant force acting between A and B_z on the considered arc. $f(z)$ is defined up to constant.

For instance, consider a uniform pressure acting on the loaded arc:

$$T_x = -pn_x = -p\frac{dy}{ds}; \quad T_y = -pn_y = p\frac{dx}{ds}$$

or

$$(T_x + iT_y)ds = p(-dy + idx) = ip(dx + idy) = ip\,dz$$

Introducing this result into [1.24], we obtain:

$$df = -p\,dz; \quad f = -pz + \text{Const} \qquad [1.25]$$

1.3. Derivation of stresses

Consider the choice $ds = dy$ in [1.22], for which $\underline{t}$ is equal to $\underline{e}_y$ so that $\underline{n}$ is equal to $\underline{e}_x$. This implies that $T_x = \sigma_{xx}$ and $T_y = \sigma_{xy}$:

$$\begin{aligned}\sigma_{xx} + i\sigma_{xy} &= -i\frac{\partial}{\partial y}\left(\phi(z) + z\overline{\phi'(z)} + \overline{\psi(z)}\right)\\ &= \phi'(z) + \overline{\phi'(z)} - z\overline{\phi''(z)} - \overline{\psi'(z)}\end{aligned} \qquad [1.26]$$

In turn, if $ds = -dx$, $\underline{t}$ is along $-\underline{e}_x$, so that $\underline{n} = \underline{e}_y$. Hence, we have $T_x = \sigma_{xy}$ and $T_y = \sigma_{yy}$:

$$\begin{aligned}\sigma_{yy} - i\sigma_{xy} &= \frac{\partial}{\partial x}\left(\phi(z) + z\overline{\phi'(z)} + \overline{\psi(z)}\right)\\ &= \phi'(z) + \overline{\phi'(z)} + z\overline{\phi''(z)} + \overline{\psi'(z)}\end{aligned} \qquad [1.27]$$

Combinations of these relations successively yield:

$$\sigma_{xx} + \sigma_{yy} = 2\left(\phi'(z) + \overline{\phi'(z)}\right) = P \qquad [1.28]$$

where [1.16] has been used, and

$$\sigma_{yy} - \sigma_{xx} + 2i\sigma_{xy} = 2\left(\psi'(z) + \overline{z}\phi''(z)\right) \qquad [1.29]$$

The stress components in cartesian and polar coordinates being related by:

$$\begin{aligned} &\sigma_{rr} + \sigma_{\theta\theta} = \sigma_{xx} + \sigma_{yy} \\ &\sigma_{\theta\theta} - \sigma_{rr} + 2i\sigma_{r\theta} = e^{2i\theta}(\sigma_{yy} - \sigma_{xx} + 2i\sigma_{xy}) \end{aligned} \qquad [1.30]$$

it is readily seen from [1.28] and [1.29] that:

$$\begin{aligned} &\sigma_{rr} + \sigma_{\theta\theta} = 2\left(\phi'(z) + \overline{\phi'(z)}\right) \\ &\sigma_{\theta\theta} - \sigma_{rr} + 2i\sigma_{r\theta} = 2e^{2i\theta}\left(\psi'(z) + \overline{z}\phi''(z)\right) \end{aligned} \qquad [1.31]$$

The stresses are not modified if $\phi(z)$ is replaced by $\phi(z) + iCz + \gamma$ and if $\psi(z)$ is replaced by $\psi(z) + \gamma'$, where γ and γ' are complex-valued constants and C is a real-valued constant. Let us assume that the origin $z = 0$ is part of the domain of study. If the boundary conditions prescribe stresses only, the arbitrariness of the definition of $\phi(z)$ and $\psi(z)$ allows us to choose them in such a way that:

$$\phi(0) = 0; \quad \psi(0) = 0; \quad \operatorname{Im}\phi'(0) = 0 \qquad [1.32]$$

When the domain of study is infinite, another possibility is to define $\phi(z)$ and $\psi(z)$ by conditions at infinity of the form:

$$\phi(\infty) = 0; \quad \psi(\infty) = 0; \quad \operatorname{Im}\phi'(\infty) = 0 \qquad [1.33]$$

1.4. Derivation of displacements

In plane strains, the isotropic linear elastic constitutive equation reads:

$$2\mu\frac{\partial\xi_x}{\partial x} = \sigma_{xx} - \nu\left(\sigma_{xx} + \sigma_{yy}\right)$$
$$2\mu\frac{\partial\xi_y}{\partial y} = \sigma_{yy} - \nu\left(\sigma_{xx} + \sigma_{yy}\right) \qquad [1.34]$$

Observing that $\sigma_{xx} = \partial^2 U/\partial y^2 = P - \partial^2 U/\partial x^2$, and using [1.16] together with [1.28], we obtain:

$$2\mu\frac{\partial\xi_x}{\partial x} = P(1-\nu) - \frac{\partial^2 U}{\partial x^2} \qquad [1.35]$$

which can be integrated in the form (see [1.16]):

$$2\mu\xi_x = 4(1-\nu)p - \frac{\partial U}{\partial x} \qquad [1.36]$$

We recall that the partial derivatives of U have been determined previously (see equations [1.18] and [1.19]).

Similarly, note that $\sigma_{yy} = \partial^2 U/\partial x^2 = P - \partial^2 U/\partial y^2$. Again, we use [1.16] and [1.28], which yields:

$$2\mu\frac{\partial\xi_y}{\partial y} = P(1-\nu) - \frac{\partial^2 U}{\partial y^2} \qquad [1.37]$$

A primitive of [1.37] reads:

$$2\mu\xi_y = 4(1-\nu)q - \frac{\partial U}{\partial y} \qquad [1.38]$$

Equations [1.36] and [1.38] define the displacement up to a rigid body motion. Finally, a combination of these equations together with [1.20] gives:

$$2\mu(\xi_x + i\xi_y) = \kappa\phi(z) - z\overline{\phi'(z)} - \overline{\psi(z)} \qquad [1.39]$$

where $\kappa = 3 - 4\nu$.

1.5. General form of the potentials ϕ and ψ

Considering applications, the domain of study S is the complex plane, except a bounded region with closed contour L. Therefore, the studied domain is non-simply connected. We aim to determine the general form of the complex-valued functions ϕ and ψ. Without loss of generality, it can be assumed that the point $z = 0$ is located within the region bounded by L, that is $z = 0 \notin S$.

Owing to [1.28], we first note that the real part of $\phi'(z)$ is single-valued, but this is possibly not the case for the imaginary part. Therefore, the integral of $\phi'(z)$ on a closed contour surrounding L is a priori not 0 and denoted by $2i\pi A$ ($A \in \mathbb{R}$). There exists a single-valued holomorphic function $F(z)$ defined on S such that:

$$\phi'(z) = A\log(z) + F(z)$$

By integration, we obtain:

$$\phi(z) = A(z\log(z) - z) + \mathcal{F}(z) \quad \text{with} \quad \mathcal{F}(z) = \int_{z_o}^{z} F(u)\, du$$

where z_o is some fixed point in S. Again, if $\mathcal{F}(z)$ is not single-valued, there exists a complex-valued constant B such that $\mathcal{F}(z) - B\log(z)$ is single-valued:

$$\phi(z) = Az\log(z) + B\log(z) + \phi^*(z) \qquad [1.40]$$

where $\phi^*(z)$ is a single-valued holomorphic function defined on S. A similar reasoning starting from [1.29] shows that there exists a complex-valued constant C such that:

$$\psi(z) = C \log(z) + \psi^*(z) \qquad [1.41]$$

where $\psi^*(z)$ is a single-valued holomorphic function defined on S.

We now recall [1.39], and take advantage of the fact that the displacement is single-valued. An anticlockwise integration around L yields:

$$2\mu[\xi_x + i\xi_y]_L = 2i\pi\left(Az(\kappa+1) + B\kappa + \overline{C}\right)$$

from which the following identities are derived:

$$A = 0; \quad B\kappa + \overline{C} = 0 \qquad [1.42]$$

We now apply [1.23] to the whole contour L:

$$F_x + iF_y = -i[\phi(z) + z\overline{\phi'(z)} + \overline{\psi(z)}]_L \qquad [1.43]$$

where F_x and F_y denote the components of the resultant force acting on the contour. In order for the unit normal $\underline{n}$ to point outward with respect to S, note that the contour must be oriented clockwise. Using [1.40], [1.41] and [1.42], we find that:

$$F_x + iF_y = 2\pi\left(\overline{C} - B\right)$$

Eventually, combining this result with [1.42], $\phi(z)$ and $\psi(z)$ take the form:

$$\phi(z) = -\frac{F_x + iF_y}{2\pi(1+\kappa)}\log(z) + \phi^*(z)$$
$$\psi(z) = \frac{\kappa(F_x - iF_y)}{2\pi(1+\kappa)}\log(z) + \psi^*(z) \qquad [1.44]$$

Let us finally add the assumption that the stresses are bounded at infinity. This being the case, consider the Laurent series expansions of $\phi^*(z)$ and $\psi^*(z)$ in S:

$$\phi^*(z) = \sum_{-\infty}^{+\infty} a_n z^n; \quad \psi^*(z) = \sum_{-\infty}^{+\infty} b_n z^n$$

We can easily see that [1.28] requires $a_n = 0$ for $n \geq 2$. In the same line of reasoning, [1.29] requires $b_n = 0$ for $n \geq 2$. It is therefore possible to put $\phi(z)$ and $\psi(z)$ in the form:

$$\phi(z) = -\frac{F_x + iF_y}{2\pi(1+\kappa)}\log(z) + \Gamma z + \phi_o(z)$$
$$\psi(z) = \frac{\kappa(F_x - iF_y)}{2\pi(1+\kappa)}\log(z) + \Gamma' z + \psi_o(z) \qquad [1.45]$$

where $\Gamma = \alpha + i\beta$ and $\Gamma' = \alpha' + i\beta'$ are complex-valued constants, and $\phi_o(z)$ and $\psi_o(z)$ being single-valued holomorphic (including the point at infinity) functions defined on S. This means that they can be put in the form (no strictly positive power in the series expansion):

$$\phi_o(z) = \sum_{-\infty}^{0} a_n z^n; \quad \psi_o(z) = \sum_{-\infty}^{0} b_n z^n \qquad [1.46]$$

In the case of stress boundary conditions, [1.33] allows us to choose $\beta = 0$ as well as $a_o = b_o = 0$, so that:

$$\phi_o(z) = \sum_{-\infty}^{-1} a_n z^n; \quad \psi_o(z) = \sum_{-\infty}^{-1} b_n z^n \qquad [1.47]$$

We still have to interpret $\Gamma = \alpha$ and $\Gamma' = \alpha' + i\beta'$. Introducing [1.45] into [1.28] and [1.29], and considering the limit $|z| \to \infty$, we obtain:

$$\sigma_{xx}^{\infty} = 2\alpha - \alpha'; \quad \sigma_{yy}^{\infty} = 2\alpha + \alpha'; \quad \sigma_{xy}^{\infty} = \beta' \qquad [1.48]$$

or:

$$\Gamma = \frac{1}{4}\left(\sigma_{xx}^{\infty} + \sigma_{yy}^{\infty}\right); \quad \Gamma' = \frac{1}{2}\left(\sigma_{yy}^{\infty} - \sigma_{xx}^{\infty}\right) + i\sigma_{xy}^{\infty} \qquad [1.49]$$

1.6. Examples

For illustrative purpose, two examples are now briefly presented.

1.6.1. *Circular cavity under pressure*

Consider an infinite domain with a circular cavity (radius R) subjected to a uniform internal pressure p. The stresses at infinity are equal to 0. Since the resulting force of the stresses acting on the cavity wall is 0, [1.45] takes on the form:

$$\phi(z) = \sum_{-\infty}^{-1} a_n z^n; \quad \psi(z) = \sum_{-\infty}^{-1} b_n z^n$$

Combining [1.24] and [1.25] yields:

$$|z| = R: \qquad \phi(z) + z\overline{\phi'(z)} + \overline{\psi(z)} = -pz$$

which also reads:

$$|z| = R: \quad pz + \sum_{-\infty}^{-1} a_n z^n + \sum_{-\infty}^{-1} n\bar{a}_n R^{2(n-1)} z^{2-n} + \sum_{-\infty}^{-1} \bar{b}_n R^{2n} z^{-n} = 0$$

In which we have replaced $\bar{z}$ by R^2/z for the points on the circle with radius R. This implies that $\overline{b_{-1}} R^{-2} = -p$. All the other coefficients (a_n or b_n) are equal to 0:

$$\phi(z) = 0; \quad \psi(z) = -p\frac{R^2}{z}$$

In polar coordinates, the stresses are given by:

$$\sigma_{rr} + \sigma_{\theta\theta} = 0$$

$$\sigma_{\theta\theta} - \sigma_{rr} + 2i\sigma_{r\theta} = 2e^{2i\theta} p\frac{R^2}{z^2}$$

which yields:

$$\sigma_{\theta\theta} = -\sigma_{rr} = p\frac{R^2}{r^2}; \quad \sigma_{r\theta} = 0$$

1.6.2. *Circular cavity in a plane subjected to uniaxial traction at infinity*

As in the previous section, the domain S is infinite with a circular cavity centered at the origin (radius R). The stresses at infinity are defined by the tensor

$$\boldsymbol{\sigma}^{\infty} = p\underline{e}_y \otimes \underline{e}_y$$

The cavity wall is free of stress. Using [1.48], we obtain:

$$2\alpha - \alpha' = 0; \quad 2\alpha + \alpha' = p; \quad \beta' = 0$$

which yields:

$$\Gamma = \frac{p}{4}; \quad \Gamma' = \frac{p}{2}$$

$$\phi(z) = \frac{p}{4}z + \sum_{-\infty}^{-1} a_n z^n; \quad \psi(z) = \frac{p}{2}z + \sum_{-\infty}^{-1} b_n z^n$$

These expressions are introduced in the boundary condition [1.24]:

$$|z| = R: \quad \frac{p}{2}z + \sum_{-\infty}^{-1} a_n z^n + \sum_{-\infty}^{-1} n\bar{a}_n R^{2(n-1)} z^{2-n} + \frac{p}{2}\frac{R^2}{z} + \sum_{-\infty}^{-1} \bar{b}_n R^{2n} z^{-n} = 0$$

As in the previous example, the coefficients of the above power series must be equal to 0. This implies that:

$$a_{-1} = \bar{b}_{-1} = -\frac{p}{2}R^2; \quad \bar{b}_{-3} = a_{-1}R^2$$

All the other a_n and b_n are equal to 0. The corresponding functions $\phi(z)$ and $\psi(z)$ are:

$$\phi(z) = \frac{p}{4}z - \frac{p}{2}\frac{R^2}{z}; \quad \psi(z) = \frac{p}{2}z - \frac{p}{2}\frac{R^2}{z} - \frac{p}{2}\frac{R^4}{z^3}$$

from which the stress field can be derived.

1.7. Conformal mapping

Conformal mapping (see, e.g. [MUS 53]) is introduced with a view to solve the problem of the elliptic hole, which corresponds to the so-called inhomogeneity model of a crack in the plane strain framework.

1.7.1. *Application of conformal mapping to plane elasticity problems*

Consider a function ω defined on the domain Σ of the complex plane $\mathbb{C}$, valued in the codomain $S \subset \mathbb{C}$:

$$\begin{aligned} \omega : \ & \Sigma \to S \\ & \zeta \to z = \omega(\zeta) \end{aligned}$$

It is assumed that the function $\omega(\zeta)$ is holomorphic[1] on Σ and that $\omega'(\zeta) \neq 0$. This function is said to be a conformal map in the sense that it preserves the angles. More precisely, consider the complex numbers $d\zeta_1$ and $d\zeta_2$ representing two elementary vectors with origin ζ. Their images dz_1 and dz_2 are:

$$dz_i = \omega'(\zeta) d\zeta_i; \quad \text{Arg}(dz_i) = \text{Arg}(\omega'(\zeta)) + \text{Arg}(d\zeta_i)$$

It follows that

$$\text{Arg}(dz_2) - \text{Arg}(dz_1) = \text{Arg}(d\zeta_2) - \text{Arg}(d\zeta_1)$$

If the domain Σ is bounded while the codomain S is infinite, it implies that $z = \omega(\zeta)$ approaches infinity in the neigborhood

1 Except possibly at a pole in the case of an infinite codomain S.

of some point of Σ. For instance, $z = \infty$ is the image of $\zeta = 0$ when $\omega(\zeta)$ is of the form:

$$\omega(\zeta) = \frac{C}{\zeta} + \text{holomorphic function on } \Sigma \qquad [1.50]$$

Accordingly, $\omega(\zeta) = 1/\zeta$ provides a conformal transform from the circular domain Σ, defined by $|\zeta| < 1$ centered at $\zeta = 0$ with radius 1 on the infinite codomain S defined by $|\zeta| > 1$.

Let S be the domain on which the mechanical problem is defined. The location in the complex plane is z. The preimage Σ of S by the conformal transformation $z = \omega(\zeta)$ is defined on the ζ-plane and valued in the z-plane. Let $\phi_1(z)$ and $\psi_1(z)$ denote the complex potentials of the solution sought in the z-plane. Let the functions $\phi(\zeta)$ and $\psi(\zeta)$ be defined on the ζ-plane by the change in variable $z = \omega(\zeta)$:

$$\phi(\zeta) = \phi_1(\omega(\zeta)); \quad \psi(\zeta) = \psi_1(\omega(\zeta)) \qquad [1.51]$$

Differentiating the definition [1.51] of ϕ, we obtain:

$$\phi_1'(z) = \frac{\phi'(\zeta)}{\omega'(\zeta)} \qquad [1.52]$$

It is assumed that the problem is defined on the z-plane by a loading of the type [1.24] on a contour L_z, in which $f(z)$ is given. L_ζ denoting the preimage $\omega^{-1}(L_z)$, the change in variables yields:

$$(\forall\zeta \in L_\zeta = \omega^{-1}(L_z)) \;\; f(\omega(\zeta)) = \phi(\zeta) + \frac{\omega(\zeta)}{\overline{\omega'(\zeta)}}\overline{\phi'(\zeta)} + \overline{\psi(\zeta)} \qquad [1.53]$$

The unknowns are the functions $\phi(\zeta)$ and $\psi(\zeta)$ that should be determined from the boundary condition [1.53]. The potentials of the initial problem are then retrieved from [1.51].

In particular, an important a case occurs when the domain of study S is the complex plane with a hole bounded by the closed contour L_z. This is the subject of the following.

1.7.2. *The domain* Σ *is the unit disc* $|\zeta| \leq 1$

From now on, Σ^+ is the unit disc $|\zeta| \leq 1$. Let us assume that we know a conformal transformation $z = \omega(\zeta)$ defined on Σ^+ such that $S = \omega(\Sigma^+)$ (Figure 1.1). We have seen that ω is of the form [1.50]. The contour L_ζ is the circle γ ($|\zeta| = 1$). In this section, we aim to show that [1.53] provides a functional equation for the unknown $\phi(\zeta)$. The other unknown $\psi(\zeta)$ can then be explicitly derived from the solution to the latter. The loading considered herein is defined by stress boundary conditions. Owing to [1.33] written for $\phi_1(z)$ and $\psi_1(z)$ at infinity, and observing that $\zeta = 0$ is the preimage of $z = \infty$ by a transformation ω of the form [1.50], it is convenient to introduce the corresponding conditions at $\zeta = 0$:

$$\phi(0) = 0; \quad \psi(0) = 0 \qquad [1.54]$$

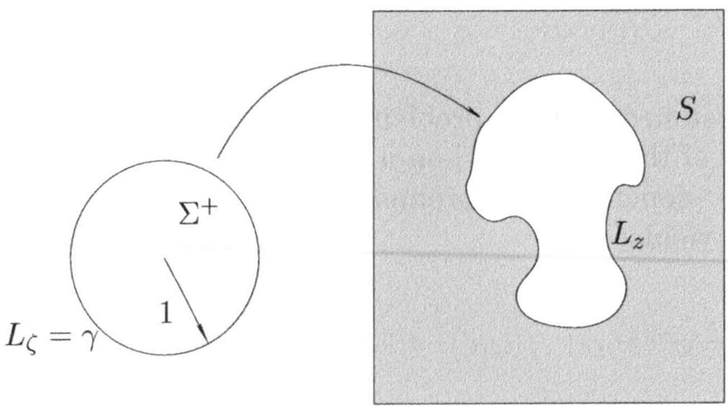

Figure 1.1. *The domain* S *mapped on the disc* $|\zeta| < 1$

To begin with, the resultant force acting on L_z is equal to 0 while the stresses tend to 0 at infinity ($\sigma^\infty = 0$). It follows that functions $\phi_1(z)$ and $\psi_1(z)$ are holomorphic on S, including the point $z = \infty$ [2]. This implies that the functions $\phi(\zeta)$ and $\psi(\zeta)$ are holomorphic in Σ^+.

Let us start with the determination of $\psi(\zeta)$ in the domain $|\zeta| < 1$ under the assumption that $\phi(\zeta)$ has been determined. Since $\psi(\zeta)$ is holomorphic in the domain $|\zeta| \leq 1$, Cauchy's formula reads:

$$|\zeta| < 1 \qquad \psi(\zeta) = \frac{1}{2i\pi} \int_\gamma \frac{\psi(\sigma)}{\sigma - \zeta} d\sigma \qquad [1.55]$$

Introducing the expression of $\psi(\sigma)$ derived from [1.53] into [1.55] yields:

$$|\zeta| < 1 \quad \psi(\zeta) = \frac{1}{2i\pi} \int_\gamma \frac{\overline{f} d\sigma}{\sigma - \zeta} - \frac{1}{2i\pi} \int_\gamma \frac{\overline{\omega(\sigma)}}{\omega'(\sigma)} \frac{\phi'(\sigma) d\sigma}{\sigma - \zeta} \qquad [1.56]$$

where formula [1.93] of Appendix 1.9 has been used as well as the condition $\overline{\phi(0)} = 0$ (see equation [1.54]); [1.56] states that $\psi(\zeta)$ can be determined in Σ^+ provided that $\phi'(\sigma)$ is known on the edge γ of the disc.

Let us now move to the determination of $\phi(\zeta)$ within the unit disc. The starting point is the theorem [1.93] written for $\psi(\zeta)$ and combined with [1.54]. We obtain:

$$|\zeta| < 1 : \qquad \frac{1}{2i\pi} \int_\gamma \frac{\overline{\psi(\sigma)}}{\sigma - \zeta} d\sigma = \overline{\psi(0)} = 0 \qquad [1.57]$$

2 This means that these functions can be expanded in a power series of the form $\sum_{k \geq 0} a_k z^{-k}$ in the neighborhood of infinity.

Then, the value $\psi(\sigma)$ of $\psi(\zeta)$ on the edge γ of the disc is taken from [1.53]:

$$|\sigma| = 1 \qquad \overline{\psi(\sigma)} = f(\omega(\sigma)) - \phi(\sigma) - \frac{\omega(\sigma)}{\overline{\omega'(\sigma)}}\overline{\phi'(\sigma)} \qquad [1.58]$$

Introducing this value into [1.57] yields:

$$\forall\zeta \text{ such that } |\zeta| < 1: \quad \phi(\zeta) + \frac{1}{2i\pi}\int_\gamma \frac{\omega(\sigma)}{\overline{\omega'(\sigma)}}\frac{\overline{\phi'(\sigma)}}{\sigma - \zeta}\,d\sigma = \frac{1}{2i\pi}\int_\gamma \frac{f\,d\sigma}{\sigma - \zeta} \qquad [1.59]$$

where Cauchy's formula:

$$|\zeta| < 1 \qquad \phi(\zeta) = \frac{1}{2i\pi}\int_\gamma \frac{\phi(\sigma)}{\sigma - \zeta}\,d\sigma \qquad [1.60]$$

has been used; [1.59] is a functional equation with respect to the unknown $\phi(\zeta)$.

Let us now consider the situation when the asymptotic stress state $\boldsymbol{\sigma}^\infty$ and the resultant force acting on the contour L_z (with components F_x and F_y) are possibly not equal to 0. The general form of functions $\phi_1(z)$ and $\psi_1(z)$ is provided by [1.45]. Recalling the form [1.50] of the conformal transformation defined on Σ^+, this implies that $\phi(\zeta)$ and $\psi(\zeta)$ are of the following type:

$$\phi(\zeta) = \frac{F_x + iF_y}{2\pi(1+\kappa)}\log(\zeta) + \frac{\Gamma C}{\zeta} + \phi_o(\zeta)$$

$$\psi(\zeta) = -\frac{\kappa(F_x - iF_y)}{2\pi(1+\kappa)}\log(\zeta) + \frac{\Gamma' C}{\zeta} + \psi_o(\zeta) \qquad [1.61]$$

where $\phi_o(\zeta)$ and $\psi_o(\zeta)$ are holomorphic on Σ^+. However, since $\phi(\zeta)$ and $\psi(\zeta)$ are not holomorphic in $\zeta = 0$, the identities [1.56] and [1.59] are no longer valid and the reasoning which has led to them must be modified. The idea consists of introducing [1.61] into [1.58], which now takes the form:

$$|\sigma| = 1 \qquad \overline{\psi_o(\sigma)} = f_o - \phi_o(\sigma) - \frac{\omega(\sigma)}{\overline{\omega'(\sigma)}}\overline{\phi_o'(\sigma)} \qquad [1.62]$$

in which f was replaced by f_o, defined by:

$$f_o = f - \frac{F_x + iF_y}{2\pi}\log\sigma - \frac{\Gamma C}{\sigma} - \frac{\omega(\sigma)}{\overline{\omega'(\sigma)}}\left(\frac{F_x - iF_y}{2\pi(1+\kappa)}\sigma - \Gamma\overline{C}\sigma^2\right) - \overline{\Gamma}'\overline{C}\sigma \qquad [1.63]$$

Since the functions ϕ_o and ψ_o are holomorphic, they can be derived from [1.56] and [1.59] in which the following changes have to be made:

$$f \to f_o, \quad \phi \to \phi_o, \quad \psi \to \psi_o$$

1.7.3. *The domain Σ is the complement Σ^- of the unit disc*

From now on, Σ^- denotes the complement of the unit disc. Hence, $\zeta \in \Sigma^-$ is equivalent to $|\zeta| \geq 1$. We assume that S is mapped by a conformal transformation $z = \omega(\zeta)$ defined on Σ^-. Furthermore, the edge γ of the unit disc is the preimage of the contour L_z ($L_\zeta = \gamma$). The general form of the transformation $\omega(\zeta)$ is

$$\omega(\zeta) = R\zeta + \omega_o(\zeta) \qquad [1.64]$$

where $R \neq 0$ is a constant and $\omega_o(\zeta)$ is holomorphic in Σ^- including at infinity.

We aim to show that [1.53] provides a functional equation for the unknown $\phi(\zeta)$. The other unknown $\psi(\zeta)$ can then be explicitly derived from the solution to the latter. As discussed in section 1.7.2, we restrict to stress boundary conditions, and we start with a loading in which $F_x = F_y = \Gamma = \Gamma' = 0$ (see equation [1.45]). In other words, the functions $\phi_1(z)$ and $\psi_1(z)$ are holomorphic, including the point $z = \infty$. This holds true for the functions $\phi(\zeta)$ and $\psi(\zeta)$, which are also holomorphic, including $\zeta = \infty$.

Let us apply theorem [1.96] of Appendix 1.9 to the function $\psi(\zeta)$ in [1.53]:

$$(\forall\sigma \in \gamma = \omega^{-1}(L_z)) \quad \overline{\psi(\sigma)} = f(\omega(\sigma)) - \phi(\sigma) - \frac{\omega(\sigma)}{\overline{\omega'(\sigma)}}\overline{\phi'(\sigma)} \quad [1.65]$$

This identity is first divided by $\sigma - \zeta$ and then integrated along γ. This yields:

$$|\zeta| > 1: \quad \frac{1}{2i\pi}\int_\gamma \frac{f\,d\sigma}{\sigma - \zeta} + \phi(\zeta) - \frac{1}{2i\pi}\int_\gamma \frac{\omega(\sigma)}{\overline{\omega'(\sigma)}}\frac{\overline{\phi'(\sigma)}}{\sigma - \zeta}\,d\sigma = 0 \quad [1.66]$$

where theorem [1.94] was applied to $\phi(\zeta)$ and the relation $\phi(\infty) = 0$ was used (see equation [1.33]); [1.66] constitutes the functional equation with respect to the unknown $\phi(\zeta)$. It is the counterpart of [1.59] for the domain $\Sigma = \Sigma^-$.

We still have to determine $\psi(\zeta)$. To do so, theorem [1.94] is applied together with the condition $\psi(\infty) = 0$:

$$|\zeta| > 1: \qquad \psi(\zeta) = -\frac{1}{2i\pi}\int_\gamma \frac{\psi(\sigma)}{\sigma - \zeta}\,d\sigma \qquad [1.67]$$

Hence, the value of $\psi(\sigma)$ follows from [1.65]:

$$(\forall\sigma \in \gamma = \omega^{-1}(L_z)) \quad \psi(\sigma) = \overline{f(\omega(\sigma))} - \overline{\phi(\sigma)} - \frac{\overline{\omega(\sigma)}}{\omega'(\sigma)}\phi'(\sigma) \quad [1.68]$$

Again using theorem [1.96], applied to $\phi(\zeta)$, it appears that:

$$|\zeta| > 1: \ \psi(\zeta) = -\frac{1}{2i\pi}\int_\gamma \frac{\overline{f}}{\sigma - \zeta}\,d\sigma + \frac{1}{2i\pi}\int_\gamma \frac{\overline{\omega(\sigma)}}{\omega'(\sigma)}\frac{\phi'(\sigma)}{\sigma - \zeta}\,d\sigma \quad [1.69]$$

We now consider the situation when the resultant force acting on $\mathcal{L}_z$ and the asymptotic stress state are possibly not equal to 0.

We start with the expressions $\phi_1(z)$ and $\psi_1(z)$ given in [1.45]. Owing to [1.51] and recalling the expression [1.64] of $\omega(\zeta)$, the potentials $\phi(\zeta)$ and $\psi(\zeta)$ defined on Σ^- are found in the form:

$$\begin{aligned}\phi(\zeta) &= -\frac{F_x + iF_y}{2\pi(1+\kappa)}\log(\zeta) + \Gamma R\zeta + \phi_o(\zeta) \\ \psi(\zeta) &= \frac{\kappa(F_x - iF_y)}{2\pi(1+\kappa)}\log(\zeta) + \Gamma' R\zeta + \psi_o(\zeta)\end{aligned} \quad [1.70]$$

where $\phi_o(\zeta)$ and $\psi_o(\zeta)$ are holomorphic on Σ^- (including the point at infinity).

We now revisit the previous reasoning starting from equation [1.65]: this equation is written in terms of $\phi_o(\zeta)$ and $\psi_o(\zeta)$, and replaces f by a new function f_o defined on γ by:

$$\begin{aligned}f_o = f - \frac{\overline{\Gamma' R}}{\sigma} - \Gamma R\left(\sigma + \frac{\omega(\sigma)}{\overline{\omega'(\sigma)}}\right) \\ + \frac{F_x - iF_y}{2\pi(1+\kappa)}\sigma\frac{\omega(\sigma)}{\overline{\omega'(\sigma)}} + \frac{F_x + iF_y}{2\pi}\log\sigma\end{aligned} \quad [1.71]$$

It is now possible to determine $\phi_o(\zeta)$ and $\psi_o(\zeta)$ from [1.66]:

$$|\zeta| > 1\text{: } \frac{1}{2i\pi}\int_\gamma \frac{f_o d\sigma}{\sigma - \zeta} + \phi_o(\zeta) - \frac{1}{2i\pi}\int_\gamma \frac{\omega(\sigma)}{\overline{\omega'(\sigma)}} \frac{\overline{\phi_o'(\sigma)}}{\sigma - \zeta} d\sigma = 0 \quad [1.72]$$

and [1.69]:

$$|\zeta| > 1\text{: } \psi_o(\zeta) = -\frac{1}{2i\pi}\int_\gamma \frac{\overline{f}_o}{\sigma - \zeta} d\sigma + \frac{1}{2i\pi}\int_\gamma \frac{\overline{\omega(\sigma)}}{\omega'(\sigma)} \frac{\phi_o'(\sigma)}{\sigma - \zeta} d\sigma \quad [1.73]$$

1.8. The anisotropic case

1.8.1. *General features*

We now consider the anisotropic case, in plane strains or plane stress conditions. It is assumed that the plane of study is a plane of material symmetry. The linear elastic constitutive equations therefore read:

$$\begin{aligned} \varepsilon_{xx} &= A_{11}\sigma_{xx} + A_{12}\sigma_{yy} + A_{16}\sigma_{xy} \\ \varepsilon_{yy} &= A_{12}\sigma_{xx} + A_{22}\sigma_{yy} + A_{26}\sigma_{xy} \\ 2\varepsilon_{xy} &= A_{16}\sigma_{xx} + A_{26}\sigma_{yy} + A_{66}\sigma_{xy} \end{aligned} \quad [1.74]$$

In the following, a_{ij} (respectively, b_{ij}) will denote the coefficients A_{ij} in plane stress (respectively, plane strain) conditions. Coefficients a_{ij} are directly related to the coefficients of the tensor of compliance:

$$\begin{aligned} &a_{11} = S_{1111}; \quad a_{22} = S_{2222}; \quad a_{12} = S_{1122} \\ &a_{16} = 2S_{1112}; \quad a_{26} = 2S_{2212}; \quad a_{66} = 4S_{1212} \end{aligned} \quad [1.75]$$

The plane strain coefficients b_{ij} are related to a_{ij} according to $(i,\ j = 1,\ 2,\ 6)$:

$$b_{ij} = a_{ij} - a_{i3}\frac{a_{3j}}{a_{33}} \quad [1.76]$$

with

$$a_{31} = S_{1133}, \quad a_{32} = S_{2233}, \quad a_{33} = S_{3333}, \quad a_{36} = 2S_{3312}$$

In the isotropic case, the conditions of geometrical compatibility [1.6] imply that the Airy function U (see equation [1.1]) is biharmonic $\Delta\Delta U = 0$ (see section 1.1). Owing to [1.74], they now take the form:

$$A_{22}\frac{\partial^4 U}{\partial x^4} - 2A_{26}\frac{\partial^4 U}{\partial x^3 \partial y} + (2A_{12} + A_{66})\frac{\partial^4 U}{\partial x^2 \partial y^2} - 2A_{16}\frac{\partial^4 U}{\partial x \partial y^3} + A_{11}\frac{\partial^4 U}{\partial y^4} = 0 \qquad [1.77]$$

Observing that all terms involve a fourth–order derivative, we find solutions in the form $f(x + \mu y)$, where μ is a complex constant and $f(z)$ is a differentiable function of the (*a priori*) complex variable $z = x + \mu y$. It appears that [1.77] will be satisfied provided that μ is a root of the polynomial equation:

$$A_{22} - 2A_{26}\mu + (2A_{12} + A_{66})\mu^2 - 2A_{16}\mu^3 + A_{11}\mu^4 = 0 \qquad [1.78]$$

The latter has two pairs of conjugate roots, respectively, μ_1, $\overline{\mu}_1$ and μ_2, $\overline{\mu}_2$. By convention, the imaginary parts of μ_1 and μ_2 are positive. Among other classical relations between coefficients of [1.78], let us note in particular that:

$$\frac{A_{22}}{A_{11}} = |\mu_1|^2|\mu_2|^2 \qquad [1.79]$$

Since $U(x, y)$ is a real-valued function, we find the solution to [1.77] in the form:

$$U(x, y) = 2\mathcal{R}e\,(f_1(x + \mu_1 y) + f_2(x + \mu_2 y)) \qquad [1.80]$$

where $f_1(z_1)$ and $f_2(z_2)$ are two differentiable functions of the two distinct complex variables:

$$z_1 = x + \mu_1 y; \quad z_2 = x + \mu_2 y \qquad [1.81]$$

1.8.2. *Stresses, displacements and boundary conditions*

Following [LEK 63] (see also [SIH 65]), let us introduce the complex potentials $\phi_1(z_1) = f_1'(z_1)$ and $\phi_2(z_2) = f_2'(z_2)$. From the definition [1.1] of the Airy function, we obtain:

$$\begin{aligned} \sigma_{xx} &= 2\mathcal{R}e\left(\mu_1^2\phi_1' + \mu_2^2\phi_2'\right) \\ \sigma_{yy} &= 2\mathcal{R}e\left(\phi_1' + \phi_2'\right) \\ \sigma_{xy} &= -2\mathcal{R}e\left(\mu_1\phi_1' + \mu_2\phi_2'\right) \end{aligned} \qquad [1.82]$$

In the case of plane strains, it is sufficient to determine the components ξ_x and ξ_y of the displacement. To do so, let us put the strains given by [1.74] (with $A_{ij} = b_{ij}$) in the form:

$$\begin{aligned} \varepsilon_{xx} &= \frac{\partial \xi_x}{\partial x} = 2\mathcal{R}e\left(p_1\phi_1' + p_2\phi_2'\right) \\ \varepsilon_{yy} &= \frac{\partial \xi_y}{\partial y} = 2\mathcal{R}e\left(q_1\mu_1\phi_1' + q_2\mu_2\phi_2'\right) \end{aligned} \qquad [1.83]$$

with the following notations ($j = 1, 2$):

$$p_j = b_{11}\mu_j^2 + b_{12} - b_{16}\mu_j; \quad q_j = b_{12}\mu_j + \frac{b_{22}}{\mu_j} - b_{26} \qquad [1.84]$$

Integration gives (up to a rigid body motion):

$$\begin{aligned} \xi_x &= 2\mathcal{R}e\left(p_1\phi_1 + p_2\phi_2\right) \\ \xi_y &= 2\mathcal{R}e\left(q_1\phi_1 + q_2\phi_2\right) \end{aligned} \qquad [1.85]$$

Moreover, the derivation of the resultant force acting on an oriented arc element – such as that presented in section 1.2 – can be directly used. [1.22] provides the resultant components along $\underline{e}_x$ and $\underline{e}_y$ of this force as a function of the curvilinear abscissa:

$$F_x(s) = \int_0^s T_x\, ds = \frac{\partial U}{\partial y} + \text{Const.}$$
$$F_y(s) = \int_0^s T_y\, ds = -\frac{\partial U}{\partial x} + \text{Const.} \qquad [1.86]$$

Using [1.80] again, the conditions on a loaded edge are put in the form:

$$2\mathcal{R}e\,(\mu_1\phi_1(z_1) + \mu_2\phi_2(z_2)) = F_x + \text{Const.}$$
$$2\mathcal{R}e\,(\phi_1(z_1) + \phi_2(z_2)) = -F_y + \text{Const.} \qquad [1.87]$$

1.9. Appendix: mathematical tools

As mentioned, γ denotes the unit circle ($|\sigma| = 1$). Cauchy's classical formula states that:

$$|\zeta| > 1 : \quad \int_\gamma \frac{d\sigma}{\sigma - \zeta} = 0 \qquad [1.88]$$

whereas

$$|\zeta| < 1 : \quad \int_\gamma \frac{d\sigma}{\sigma - \zeta} = 2i\pi \qquad [1.89]$$

Furthermore, a partial fraction decomposition reads:

$$k \geq 1 : \quad \frac{1}{\sigma^k(\sigma - \zeta)} = -\sum_{j=1}^{k} \frac{1}{\zeta^j}\frac{1}{\sigma^{k+1-j}} + \frac{1}{(\sigma - \zeta)}\frac{1}{\zeta^k} \qquad [1.90]$$

It follows that:

$$k \geq 1,\ |\zeta| > 1: \quad \frac{1}{2i\pi} \int_{\gamma} \frac{d\sigma}{\sigma^k(\sigma - \zeta)} = -\frac{1}{\zeta^k} \qquad [1.91]$$

whereas

$$k \geq 1,\ |\zeta| < 1: \quad \frac{1}{2i\pi} \int_{\gamma} \frac{d\sigma}{\sigma^k(\sigma - \zeta)} = 0 \qquad [1.92]$$

These identities are helpful to establish the following results.

1.9.1. *Theorem 1*

If $f(\zeta)$ is a holomorphic function on the unit disc Σ^+, then:

$$|\zeta| < 1: \qquad \frac{1}{2i\pi} \int_{\gamma} \frac{\overline{f(\sigma)}}{\sigma - \zeta}\, d\sigma = \overline{f(0)} \qquad [1.93]$$

This result is readily established by means of a power series expansion $\sum_{k\geq 0} a_k \zeta^k$ of $f(\zeta)$. First, considering the boundary γ of the disc, we have:

$$|\sigma| = 1: \qquad \overline{f(\sigma)} = \sum_{k\geq 0} \overline{a}_k \sigma^{-k}$$

which is introduced into the left-hand side of [1.93]. The theorem then immediately follows from [1.89] and [1.92].

1.9.2. *Theorem 2*

The matter of the present theorem is Cauchy's formula for an infinite domain: If $f(z)$ is holomorphic in Σ^- ($|\zeta| \geq 1$) (including the point at infinity), then:

$$|\zeta| > 1: \qquad \frac{1}{2i\pi}\int_\gamma \frac{f(\sigma)}{\sigma - \zeta}\,d\sigma = f(\infty) - f(\zeta) \qquad [1.94]$$

This can be proved from a Laurent power series of $f(\zeta)$ (with no positive power since $f(\zeta)$ is holomorphic at infinity):

$$f(\zeta) = \sum_{k=-\infty}^{0} a_k \zeta^k \qquad [1.95]$$

A direct application of [1.91] and [1.88] yields

$$|\zeta| > 1: \quad \frac{1}{2i\pi}\int_\gamma \frac{f(\sigma)}{\sigma - \zeta}\,d\sigma = -\sum_{k=-\infty}^{-1} a_k \zeta^k$$

and the theorem [1.94] is established.

1.9.3. *Theorem 3*

The counterpart of theorem [1.93] for a holomorphic function in Σ^- (including the point at infinity) reads:

$$|\zeta| > 1: \qquad \frac{1}{2i\pi}\int_\gamma \frac{\overline{f(\sigma)}}{\sigma - \zeta}\,d\sigma = 0 \qquad [1.96]$$

Again, the proof is based on a power series expansion of $f(\zeta)$ in the form [1.95]. Considering in particular the unit circle γ $(\overline{\sigma} = 1/\sigma)$:

$$\overline{f(\sigma)} = \sum_{-\infty}^{0} \overline{a}_k \sigma^{-k}$$

Then, observing that the function σ^{-k} $(k \leq 0)$ is holomorphic on the unit disc, it is readily seen that

$$k \leq 0,\ |\zeta| > 1: \qquad \int_{\gamma} \frac{\sigma^{-k} d\sigma}{\sigma - \zeta} = 0$$

The theorem [1.96] immediately follows from the two last equations.

2

Fundamentals of Elasticity in View of Homogenization Theory

2.1. Green's function concept

Let (z_1, z_2, z_3) denote a cartesian coordinate system associated with the orthonormal basis $(\underline{e}_p)_{p=1,2,3}$. Consider the elementary displacement field solution $\underline{\xi}^{(p)}$ induced in an infinite homogeneous elastic continuum (stiffness tensor $\mathbb{C}$) by a unit point force $\underline{f}$ located at the origin and parallel to $\underline{e}_p$, i.e. $\underline{f} = \delta_0(\underline{z})\underline{e}_p$ (Kelvin problem). The Navier equation reads:

$$\operatorname{div}\left(\mathbb{C}\colon \mathbf{grad}\,\underline{\xi}^{(p)}\right) + \delta_0(\underline{z})\underline{e}_p = 0 \qquad [2.1]$$

The second-order Green tensor $\mathbf{G}$ is defined by $\mathbf{G} \cdot \underline{e}_p = \underline{\xi}^{(p)}$. Its use in [2.1] entails the following conditions to be satisfied by the components of $\mathbf{G}$:

$$C_{ijkl}G_{kp,jl} + \delta(\underline{z})\delta_{ip} = 0 \qquad [2.2]$$

where subscript jl denotes the differential operator $\partial^2/\partial z_j \partial z_l$.

Note that G is a symmetric tensor. This property is a straightforward consequence of Maxwell–Betti reciprocity theorem.

2.2. Green's function in two-dimensional conditions

The conditions of plane strain lead to solving a two-dimensional problem in which the loading is defined by a distribution of forces:

$$\underline{f}(z_1, z_2, z_3) = \delta(z_1, z_2)\underline{e}_p \quad (p = 1,\, 2)$$

where $\delta(z_1, z_2)$ denotes the Dirac function for the two-dimensional space. Accordingly, the displacement is sought as a vector field in the plane defined by $\underline{e}_1$ and $\underline{e}_2$ (no component along $\underline{e}_3$).

2.2.1. *The general anisotropic case*

The starting point is the plane-wave decomposition of the Dirac distribution for the two-dimensional space, in the form:

$$\delta(\underline{z}) = -\frac{1}{4\pi^2}\int_{|\underline{\eta}|=1} \frac{d\ell}{(\underline{\eta}\cdot\underline{z})^2} \qquad [2.3]$$

in which the integral is performed on the unit circle ($|\underline{\eta}| = 1$). Introducing this expression into [2.2] yields:

$$C_{ijkl}G_{kp,jl} - \frac{\delta_{ip}}{4\pi^2}\int_{|\underline{\eta}|=1} \frac{d\ell}{(\underline{\eta}\cdot\underline{z})^2} = 0 \qquad [2.4]$$

Taking advantage of the linearity of this equation with respect to G_{kp}, the idea is to consider a given value of $\underline{\eta}$ on the unit circle and to search for the solution $\mathbf{G}^{\eta}(\underline{z})$ of:

$$C_{ijkl}G^{\eta}_{kp,jl} - \frac{\delta_{ip}}{4\pi^2(\underline{\eta}\cdot\underline{z})^2} = 0 \qquad [2.5]$$

We look for $\mathbf{G}^{\eta}(\underline{z})$ in the form $g(\underline{\eta}\cdot\underline{z})\mathbf{H}(\underline{\eta})$. Indeed, this yields:

$$\eta_j\eta_l C_{ijkl}H_{kp}\, g''(\underline{\eta}\cdot\underline{z}) - \frac{\delta_{ip}}{4\pi^2(\underline{\eta}\cdot\underline{z})^2} = 0 \qquad [2.6]$$

It is useful to introduce the so-called acoustic tensor

$$\mathbf{K}(\underline{\eta}) = \underline{\eta}\cdot\mathbb{C}\cdot\underline{\eta}$$

with components $K_{ik}(\underline{\eta}) = \eta_j\eta_l C_{ijkl}$. With this notation, it is readily seen that [2.6] takes the form:

$$\mathbf{K}(\underline{\eta})\cdot\mathbf{H}(\underline{\eta})g''(\zeta) + \frac{1}{4\pi^2}\frac{d^2}{d\zeta^2}\left(\log(\zeta)\right)\mathbf{1} = 0 \quad \text{with} \quad \zeta = \underline{\eta}\cdot\underline{z} \qquad [2.7]$$

An immediate solution of [2.7] reads

$$\mathbf{G}^{\eta}(\underline{z}) = -\left(\mathbf{K}(\underline{\eta})\right)^{-1}\log(|\underline{\eta}\cdot\underline{z}|)/(4\pi^2).$$

We now return to [2.4]. By superposition, we obtain:

$$\mathbf{G}(\underline{z}) = -\frac{1}{4\pi^2}\int_{|\underline{\eta}|=1}\left(\mathbf{K}(\underline{\eta})\right)^{-1}\log(|\underline{\eta}\cdot\underline{z}|)\,d\ell \qquad [2.8]$$

2.2.2. *The isotropic case*

In the isotropic case, let k and μ denote the bulk and shear moduli. The stiffness tensor reads:

$$\mathbb{C} = 3k\mathbb{J} + 2\mu\mathbb{K} \qquad [2.9]$$

where $\mathbb{J}$ and $\mathbb{K}$ are the spherical and deviatoric isotropic fourth-order projectors[1]. The acoustic tensor takes the form:

$$\mathbf{K} = \mu \left(\frac{4\mu + 3k}{3\mu} \underline{\eta} \otimes \underline{\eta} + \mathbf{1} - \underline{\eta} \otimes \underline{\eta} \right) \qquad [2.10]$$

and its inverse is:

$$\mathbf{K}^{-1} = \frac{1}{\mu} \left(\mathbf{1} - \frac{\mu + 3k}{4\mu + 3k} \underline{\eta} \otimes \underline{\eta} \right) \qquad [2.11]$$

Introducing [2.11] into [2.8] yields:

$$\mathbf{G}(\underline{z}) = -\frac{1}{4\pi^2 \mu} \left(\mathcal{G}(\underline{z}) \mathbf{1} - \frac{\mu + 3k}{4\mu + 3k} \gamma(\underline{z}) \right) \qquad [2.12]$$

with

$$\mathcal{G}(\underline{z}) = \int_{|\underline{\eta}=1|} \log(|\underline{\eta} \cdot \underline{z}|) \, d\ell;$$

$$\gamma(\underline{z}) = \int_{|\underline{\eta}=1|} \log(|\underline{\eta} \cdot \underline{z}|) \, \underline{\eta} \otimes \underline{\eta} \, d\ell \qquad [2.13]$$

Let us first consider the scalar integral $\mathcal{G}(\underline{z})$. It is convenient to introduce the angle θ between $\underline{z}$ and $\underline{\eta}$, so that $\underline{\eta} \cdot \underline{z} = r \cos\theta$ and $r = |\underline{z}|$. Observing that $\log(|\underline{\eta} \cdot \underline{z}|) = \log r + \log|\cos\theta|$, and recalling that $d\ell = d\theta$ on the unit circle, we conclude that:

$$\mathcal{G}(\underline{z}) = 2\pi \log r + \text{Const.}, \qquad [2.14]$$

1 $J_{ijkl} = \frac{1}{3}\delta_{ij}\delta_{kl}$ and $K_{ijkl} = \frac{1}{2}(\delta_{ik}\delta_{jl} + \delta_{il}\delta_{jk}) - \frac{1}{3}\delta_{ij}\delta_{kl}$.

where the constant[2] can be omitted since we are interested in the derivatives of the Green's function.

In view of determining $\gamma(\underline{z})$, let us more generally consider any integral of the form:

$$\mathcal{X}(\underline{z}) = \int_{|\underline{\eta}=1|} X(\underline{\eta} \cdot \underline{z})\, d\ell \qquad [2.15]$$

so:

$$\frac{\partial^2 \mathcal{X}}{\partial z_i \partial z_j}(\underline{z}) = \int_{|\underline{\eta}=1|} \eta_i \eta_j X''(\underline{\eta} \cdot \underline{z})\, d\ell \qquad [2.16]$$

The coefficient $\gamma_{ij}(\underline{z})$ proves to be identical to the integral in [2.16] with $X''(\zeta) = \log \zeta$, that is $X(\zeta) = \zeta^2(\log \zeta - 3/2)/2$. We thus seek the second derivatives $\mathcal{X}_{,ij}(\underline{z})$ of the function $\mathcal{X}(\underline{z})$ of [2.15], for this choice of $X(\zeta)$:

$$\begin{aligned} \mathcal{X}(\underline{z}) &= \frac{1}{2} r^2 \log r \int_0^{2\pi} \cos^2\theta\, d\theta \\ &+ \frac{r^2}{2} \int_0^{2\pi} \cos^2\theta (\log|\cos\theta| - \frac{3}{2})\, d\theta \end{aligned} \qquad [2.17]$$

The contribution to the derivative $\mathcal{X}_{,ij}(\underline{z})$ of the term in r^2 in the above integral is a constant that can be omitted:

$$\gamma_{ij}(\underline{z}) = \frac{\pi}{2}(r^2 \log r)_{,ij} + \text{Const.} \qquad [2.18]$$

Introducing [2.14] and [2.18] into [2.12]:

$$G_{ij}(\underline{z}) = \frac{1}{2\pi\mu} \left(\frac{\mu + 3k}{4(4\mu + 3k)} (r^2 \log r)_{,ij} - \delta_{ij} \log r \right) \qquad [2.19]$$

2 Which is equal to $-2\pi \log 2$.

It can be useful to note that:

$$(r^2 \log r)_{,ij} = 2(r_{,i} r_{,j} + \delta_{ij} \log r) + \delta_{ij} \qquad [2.20]$$

Omitting the constant terms in the above equation, an alternative expression of $G_{ij}(\underline{z})$ can be derived from [2.19]:

$$G_{ij}(\underline{z}) = \frac{1}{8\pi\mu(1-\nu)} \left(r_{,i} r_{,j} - (3-4\nu)\delta_{ij} \log r\right) \qquad [2.21]$$

2.3. Green's function in three-dimensional conditions

2.3.1. *The general anisotropic case*

Following [WIL 77], the starting point for the determination of the Green's function in three-dimensional conditions is again the plane-wave decomposition of the Dirac distribution:

$$\delta(\underline{z}) = -\frac{1}{8\pi^2} \int_{|\underline{\eta}|=1} \delta''(\underline{\eta} \cdot \underline{z})\, dS_\eta \qquad [2.22]$$

where the integral is performed over the unit sphere ($|\underline{\eta}| = 1$) and dS_η is the elementary surface on the latter. The right-hand side involves the second derivative of the one-dimensional Dirac distribution.

The reasoning is now strictly identical to that presented in section 2.2.1. First, [2.22] is introduced into [2.2], so that [2.4] is replaced by:

$$C_{ijkl} G_{kp,jl} - \frac{\delta_{ip}}{8\pi^2} \int_{|\underline{\eta}|=1} \delta''(\underline{\eta} \cdot \underline{z})\, dS_\eta = 0 \qquad [2.23]$$

For a given value of $\underline{\eta}$ on the unit sphere, we then seek the solution $\mathbf{G}^{\eta}(\underline{z})$ of:

$$C_{ijkl}G^{\eta}_{kp,jl} - \frac{\delta_{ip}}{8\pi^2}\delta''(\underline{\eta}\cdot\underline{z}) = 0 \qquad [2.24]$$

which is the counterpart of [2.5]. Again, we seek $\mathbf{G}^{\eta}(\underline{z})$ in the form $g(\underline{\eta}\cdot\underline{z})\mathbf{H}(\underline{\eta})$, [2.6] and [2.7] being, respectively, replaced by:

$$\eta_j\eta_l C_{ijkl}H_{kp}\, g''(\underline{\eta}\cdot\underline{z}) - \frac{\delta_{ip}}{8\pi^2}\delta''(\underline{\eta}\cdot\underline{z}) = 0 \qquad [2.25]$$

and

$$\mathbf{K}(\underline{\eta})\cdot\mathbf{H}(\underline{\eta})g''(\zeta) - \frac{1}{8\pi^2}\frac{d^2}{d\zeta^2}\left(\delta(\zeta)\right)\mathbf{1} = 0 \quad \text{with} \quad \zeta = \underline{\eta}\cdot\underline{z} \qquad [2.26]$$

In comparison to the two-dimensional case, we just have to replace $\log\zeta$ by $-\delta(\zeta)/2$, where $\zeta = \underline{\eta}\cdot\underline{z}$, in [2.7] and [2.8]. This yields:

$$\mathbf{G}(\underline{z}) = \frac{1}{8\pi^2}\int_{|\underline{\eta}|=1}\left(\mathbf{K}(\underline{\eta})\right)^{-1}\delta(\underline{\eta}\cdot\underline{z})\,dS_\eta \qquad [2.27]$$

2.3.2. *The isotropic case*

Since [2.10] and [2.11] are still valid, the determination of the Green's function in isotropic conditions amounts to that of the two following integrals:

$$\mathcal{J} = \int_{|\underline{\eta}|=1}\delta(\underline{\eta}\cdot\underline{z})\,dS_\eta;$$

$$\mathbf{J}(\underline{z}) = \int_{|\underline{\eta}|=1}\underline{\eta}\otimes\underline{\eta}\,\delta(\underline{\eta}\cdot\underline{z})\,dS_\eta \qquad [2.28]$$

Indeed, with these notations, a combination of [2.11] and [2.27] yields:

$$\mathbf{G}(\underline{z}) = \frac{1}{8\pi^2\mu}\left(\mathcal{J}(\underline{z})\mathbf{1} - \frac{\mu+3k}{4\mu+3k}\mathbf{J}(\underline{z})\right) \qquad [2.29]$$

For the determination of $\mathbf{J}(\underline{z})$, it is interesting to introduce the three-dimensional counterpart of [2.15] and [2.16], namely:

$$\mathcal{X}(\underline{z}) = \int_{|\underline{\eta}|=1} X(\underline{\eta}\cdot\underline{z})\, dS_\eta;$$

$$\frac{\partial^2 \mathcal{X}}{\partial z_i \partial z_j}(\underline{z}) = \int_{|\underline{\eta}|=1} \eta_i\eta_j X''(\underline{\eta}\cdot\underline{z})\, dS_\eta \qquad [2.30]$$

the integrals in [2.30] now being performed over the unit sphere. We observe that $J_{ij}(\underline{z})$ is identical to $\mathcal{X}_{,ij}(\underline{z})$ provided that $X''(\zeta) = \delta(\zeta)$. Recalling that the Heaviside function[3] $\mathcal{Y}(\zeta)$ satisfies $\mathcal{Y}'(\zeta) = \delta(\zeta)$, the appropriate choice proves to be $X(\zeta) = \zeta\mathcal{Y}(\zeta)$. Recalling the notation $\underline{\eta}\cdot\underline{z} = r\cos\theta$, this yields

$$\mathcal{X}(\underline{z}) = \int_0^{2\pi} d\varphi \int_0^{\pi} r\cos\theta\,\mathcal{Y}(r\cos\theta)\sin\theta\, d\theta = \pi r \text{ with } r = |\underline{z}| \qquad [2.31]$$

and accordingly:

$$\mathbf{J}(\underline{z}) = \frac{\pi}{r}(\mathbf{1} - \underline{Z}\otimes\underline{Z}) \quad \text{with} \quad \underline{Z} = \frac{\underline{z}}{|\underline{z}|} \qquad [2.32]$$

Observing that $\mathcal{J}(\underline{z}) = \operatorname{tr}\mathbf{J}(\underline{z}) = 2\pi/r$, [2.29] eventually takes the form:

$$\mathbf{G}(\underline{z}) = \frac{1}{8\pi\mu r}\left(\frac{7\mu+3k}{4\mu+3k}\mathbf{1} + \frac{\mu+3k}{4\mu+3k}\underline{Z}\otimes\underline{Z}\right) \text{ with } \underline{Z} = \frac{\underline{z}}{|\underline{z}|} \qquad [2.33]$$

3 The Heaviside $\mathcal{Y}(\zeta)$ is defined by $\mathcal{Y}(\zeta) = 0$ for $\zeta \leq 0$, $\mathcal{Y}(\zeta) = 1$ for $\zeta > 0$.

2.4. Eshelby's problems in linear microelasticity

We successively discuss Eshelby's inclusion and inhomogeneity problems [ESH 57], which will prove very useful for the derivation of micromechanics estimates of the stiffness of cracked media (Chapters 10 and 11). The Green operator presented in section 2.4.2 will be helpful in the derivation of bounds (Chapter 12).

2.4.1. *The (elastic) inclusion problem*

In the so-called inclusion problem, an infinite medium with homogeneous stiffness tensor $\mathbb{C}$ is considered. The strain tensor ε tends toward a uniform value $\mathbf{E}^{\infty}$ at infinity, or equivalently, the displacement field $\underline{\xi}$ tends toward $\mathbf{E}^{\infty} \cdot \underline{z}$. The classical linear relationship between the stress tensor $\boldsymbol{\sigma}$ and strain tensor ε is adopted, except in a bounded domain I where a uniform initial stress $\boldsymbol{\tau}^I$ is introduced:

$$\begin{array}{ll} \operatorname{div} \boldsymbol{\sigma} = 0 & (a) \\ \boldsymbol{\sigma} = \mathbb{C} : \boldsymbol{\varepsilon} + \boldsymbol{\tau}^I \chi^I(\underline{z}) & (b) \\ \underline{\xi} = \mathbf{E}^{\infty} \cdot \underline{z} \text{ when } \underline{z} \to \infty & (c) \end{array} \qquad [2.34]$$

where χ^I is the indicator function[4] of I. We begin with the case $\mathbf{E}^{\infty} = 0$. Substituting [2.34b] into [2.34a] yields the momentum balance equation in the form:

$$\operatorname{div}\,(\mathbb{C} : \boldsymbol{\varepsilon}) + \boldsymbol{\tau}^I \cdot \boldsymbol{\nabla}\, \chi^I = 0 \qquad [2.35]$$

$\boldsymbol{\nabla}\, \chi^I$ involves the derivation of a discontinuous function across the boundary ∂I. It therefore involves the Dirac

4 $\chi^I(\underline{z}) = 0$ if $\underline{z} \notin I$; $\chi^I(\underline{z}) = 1$ if $\underline{z} \in I$.

distribution[5] $\delta_{\partial I}$ associated with the boundary ∂I of the domain I. More precisely, $\underline{n}$ denoting the unit normal vector to ∂I oriented outward with respect to I, we have $\nabla \chi^I = -\underline{n}\delta_{\partial I}$. Equation [2.35] thus becomes:

$$\operatorname{div}(\mathbb{C} : \boldsymbol{\varepsilon}) - \boldsymbol{\tau}^I \cdot \underline{n}\delta_{\partial I} = 0 \qquad [2.36]$$

The solution of [2.36] is derived using the Green's function concept. An immediate generalization of [2.2] consists of observing that the displacement induced at point $\underline{z}$ by a unit point force $\delta_{\underline{z}'}\underline{e}_p$ located at point $\underline{z}'$ reads $\mathcal{G}(\underline{z}, \underline{z}') = \mathbf{G}(\underline{z} - \underline{z}') \cdot \underline{e}_p$. Starting from this unit point-load solution, the solution to [2.36] is obtained by superposition:

$$\underline{\xi}(\underline{z}) = -\int_{\partial I} \mathbf{G}(\underline{z} - \underline{z}') \cdot \boldsymbol{\tau}^I \cdot \underline{n}(\underline{z}')\, dS_{z'} \qquad [2.37]$$

That is,

$$\xi_i(\underline{z}) = -\int_{\partial I} G_{il}(\underline{z} - \underline{z}') n_k(\underline{z}')\, dS_{z'}\, \tau^I_{lk} \qquad [2.38]$$

where the integration in [2.37] and [2.38] is performed with respect to $\underline{z}'$. Using the divergence theorem also gives:

$$\begin{aligned} \xi_i(\underline{z}) &= -\int_I \frac{\partial}{\partial z'_k} G_{il}(\underline{z} - \underline{z}')\, dV_{z'}\, \tau^I_{lk} \\ &= \frac{\partial}{\partial z_k} \left(\int_I G_{il}(\underline{z} - \underline{z}')\, dV_{z'} \right) \tau^I_{lk} \end{aligned} \qquad [2.39]$$

5 Let φ denote a C^∞ function equal to 0 outside a bounded domain. $\delta_{\partial I}$ is the distribution defined by:

$$\int \delta_{\partial I}(\underline{z})\varphi(\underline{z})\, dV = \int_{\partial I} \varphi(\underline{z})\, dS$$

Finally, the strain tensor is obtained by an additional derivation[6]:

$$\varepsilon_{ij}(\underline{z}) = \left(\frac{\partial^2}{\partial z_j \partial z_k} \left(\int_I G_{il}(\underline{z} - \underline{z}')\, dV_{z'} \right) \right)_{(ij)} \tau^I_{lk} \qquad [2.40]$$

or equivalently, making use of the symmetry of $\boldsymbol{\tau}^I$:

$$\boldsymbol{\varepsilon}(\underline{z}) = -\mathbb{P}(\underline{z}) : \boldsymbol{\tau}^I \qquad [2.41]$$

where:

$$P_{ijkl}(\underline{z}) = -\left(\frac{\partial^2}{\partial z_j \partial z_k} \left(\int_I G_{il}(\underline{z} - \underline{z}')\, dV_{z'} \right) \right)_{(ij),(kl)} \qquad [2.42]$$

By superposition, the complete solution of [2.34] as a function of $\boldsymbol{\tau}^I$ and $\mathbf{E}^\infty$ reads:

$$\boldsymbol{\varepsilon}(\underline{z}) = -\mathbb{P}(\underline{z}) : \boldsymbol{\tau}^I + \mathbf{E}^\infty \qquad [2.43]$$

With [2.43], which is based on the assumption $\boldsymbol{\tau}^I = \text{const}$, it appears that the solution of [2.34] reduces to the determination of the fourth-order $\mathbb{P}$ tensor (see sections 2.5 and 2.6), which in turn goes back to that of the second-order Green's function $\mathbf{G}(\underline{z} - \underline{z}')$ (see section 2.1).

The $\mathbb{P}$ tensor is often referred to as Hill's tensor. The essential result, which is shown in sections 2.5 and 2.6, is that Hill's tensor takes a uniform value in the domain I when the latter is an ellipsoid in three-dimensional conditions, or an ellipse in plane elasticity:

$$(\forall \underline{z} \in I) \quad \mathbb{P}(\underline{z}) = \mathbb{P}^I \qquad [2.44]$$

6 As previously, the subscript (ij) stands for a symmetrization with respect to i and j.

This implies that the strain field is uniform in an ellipsoidal (respectively, elliptic) inclusion; [2.41] then reads:

$$(\forall \underline{z} \in I) \quad \boldsymbol{\varepsilon}(\underline{z}) = -\mathbb{P}^I : \boldsymbol{\tau}^I \qquad [2.45]$$

This further implies that the skewsymmetric part of the deformation gradient in such an inclusion is also uniform, so that it undergoes a homogeneous transformation.

It may be useful to note that [2.39] also provides the deformation gradient:

$$\xi_{i,j}(\underline{z}) = \frac{\partial^2}{\partial z_j \partial z_k} \left(\int_I G_{il}(\underline{z} - \underline{z}')\, dV_{z'} \right) \tau^I_{lk} \qquad [2.46]$$

or, making use of the symmetry of $\boldsymbol{\tau}^I$:

$$\nabla \underline{\xi} = -\mathbb{R} : \boldsymbol{\tau}^I \qquad [2.47]$$

where

$$R_{ijkl}(\underline{z}) = -\left(\frac{\partial^2}{\partial z_j \partial z_k} \left(\int_I G_{il}(\underline{z} - \underline{z}')\, dV_{z'} \right) \right)_{(kl)} \qquad [2.48]$$

As opposed to [2.42], no symmetrization is done with respect to the subscripts i and j. As already stated, the tensor $\mathbb{R}$ will take a constant value $\mathbb{R}^I$ inside an ellipsoidal (respectively, elliptic) inclusion.

2.4.2. *The Green operator of the infinite space*

We consider now a generalized form of [2.34] in which the initial (or polarization) stress field $\boldsymbol{\tau}(\underline{z})$ is given:

$$\begin{array}{ll} \operatorname{div} \boldsymbol{\sigma} = 0 & (a) \\ \boldsymbol{\sigma} = \mathbb{C} : \boldsymbol{\varepsilon} + \boldsymbol{\tau}(\underline{z}) & (b) \\ \underline{\xi} = \mathbf{E}^\infty \cdot \underline{z} \text{ when } \underline{z} \to \infty & (c) \end{array} \qquad [2.49]$$

To begin with, the strain $\mathbf{E}^\infty$ at infinity is 0 and the strain field $\boldsymbol{\varepsilon}$ solution is sought as a function of the polarization field $\boldsymbol{\tau}$. The idea is to take advantage of the solution found in section 2.4.1 in the case of a uniform polarization stress in a bounded domain. More precisely, it consists of representing the polarization field $\boldsymbol{\tau}(\underline{z}')$ as an infinity of superimposed elementary uniform polarization fields, each being restricted to the elementary domains $dV_{\underline{z}'}$ centered at $\underline{z}'$. The elementary polarization of $dV_{\underline{z}'}$ by $\boldsymbol{\tau}(\underline{z}')$ induces at point $\underline{z}$ an elementary strain $d\boldsymbol{\varepsilon}(\underline{z})$ obtained from [2.41] and [2.42]. The domain I in [2.42] is $dV_{\underline{z}'}$, so that the integral sign can be removed and $\boldsymbol{\tau}^I$ in [2.41] is replaced by $\boldsymbol{\tau}(\underline{z}')$:

$$\underline{z}' \neq \underline{z} : \qquad d\varepsilon_{ij}(\underline{z}) = \frac{\partial^2 G_{ik}}{\partial z_l \partial z_j}(\underline{z} - \underline{z}')\tau_{kl}(\underline{z}')dV_{\underline{z}'} \Big|_{(ij)} \qquad [2.50]$$

It is convenient to introduce the tensor $\mathbb{\Gamma}^\infty(\underline{z})$, the general term of which is defined by

$$\Gamma^\infty_{ijkl}(\underline{z}) = -\frac{\partial^2 G_{ik}}{\partial z_l \partial z_j}\Big|_{(ij)(kl)}(\underline{z}) \qquad [2.51]$$

The symmetry of the tensor $\mathbb{\Gamma}^\infty(\underline{z})$ is an important property, which is a direct consequence of the symmetry of the Green function $\mathbf{G}$.

With this notation, [2.50] takes the form:

$$d\boldsymbol{\varepsilon}(\underline{z}) = -\mathbb{\Gamma}^\infty(\underline{z} - \underline{z}') : \boldsymbol{\tau}(\underline{z}')dV_{z'} \qquad [2.52]$$

Equation [2.52] is valid provided that $\underline{z}' \neq \underline{z}$. In turn, we must consider the effect at point $\underline{z}$ of the uniform polarization of an elementary domain dV_z centered at point $\underline{z}$ by the polarization stress $\boldsymbol{\tau}(\underline{z})$. For instance, let us define dV_z as the infinitesimal sphere $B(\underline{z}, \epsilon)$ centered at $\underline{z}$, with radius ϵ

($\epsilon \to 0$): The strain induced at point $\underline{z}$ by $\boldsymbol{\tau}(\underline{z})$ is derived from [2.45] with $I = B(\underline{z}, \epsilon)$ and $\boldsymbol{\tau}^I = \boldsymbol{\tau}(\underline{z})$, which yields $-\mathbb{P}^{\text{sph}} : \boldsymbol{\tau}(\underline{z})$. In this expression, $\mathbb{P}^{\text{sph}}$ is the Hill tensor of a spherical inclusion. For instance, in the isotropic case, $\mathbb{P}^{\text{sph}}$ can be derived from [2.33] and [2.42] and reads:

$$\mathbb{P}^{\text{sph}} = \frac{\alpha}{3k}\mathbb{J} + \frac{\beta}{2\mu}\mathbb{K} \text{ where } \alpha = \frac{3k}{3k + 4\mu}; \; \beta = \frac{6}{5}\frac{k + 2\mu}{3k + 4\mu} \qquad [2.53]$$

Eventually, the strain $\boldsymbol{\varepsilon}(\underline{z})$ induced at point $\underline{z}$ by the polarization field $\boldsymbol{\tau}(\underline{z}')$ is the sum of the contributions of the form [2.52] (for $\underline{z}' \neq \underline{z}$) and of the singular contribution of point $\underline{z}$. It appears that the strain $\boldsymbol{\varepsilon}(\underline{z})$ can be put in the form $\boldsymbol{\varepsilon}(\underline{z}) = -(\Gamma^\infty * \tau)(\underline{z})$ where Γ^∞ is a linear operator acting on the polarization field $\tau(\underline{z}')$ according to:

$$-(\Gamma^\infty * \tau)(\underline{z}) = -P.V. \int \mathbb{\Gamma}^\infty(\underline{z} - \underline{z}') : \boldsymbol{\tau}(\underline{z}')\, dV_{z'} - \mathbb{P}^{\text{sph}} : \boldsymbol{\tau}(\underline{z}) [2.54]$$

In [2.54], the Cauchy principal value ($P.V.$) of the integral on the right-hand side is defined as:

$$P.V. \int \mathbb{\Gamma}^\infty(\underline{z} - \underline{z}') : \boldsymbol{\tau}(\underline{z}')\, dV_{z'} = \lim_{\epsilon \to 0} \int_{\mathbb{R}^3 - B(\underline{z}, \epsilon)} \mathbb{\Gamma}^\infty(\underline{z} - \underline{z}') : \boldsymbol{\tau}(\underline{z}')\, dV_{\underline{z}'} \qquad [2.55]$$

For the sake of conciseness, the following notation will be used:

$$\Gamma^\infty * \boldsymbol{\tau}(\underline{z}) = \int \mathbb{\Gamma}^\infty(\underline{z} - \underline{z}') : \boldsymbol{\tau}(\underline{z}')\, dV_{z'} = P.V. \int \mathbb{\Gamma}^\infty(\underline{z} - \underline{z}') : \boldsymbol{\tau}(\underline{z}')\, dV_{z'} + \mathbb{P}^{sph} : \boldsymbol{\tau}(\underline{z}) \qquad [2.56]$$

The expression of $\Gamma^\infty * \boldsymbol{\tau}$ involves a remote part and a local part. The first term is the remote part and is the Cauchy P.V. of an integral: it consists of integrating with respect to $\underline{z}'$, except a sphere centered at $\underline{z}$ with radius $\epsilon \to 0$. The second term in the right-hand side of [2.54], called singular (or local) part, represents the self-influence at point $\underline{z}$. The Hill tensor $\mathbb{P}^{\text{sph}}$ is associated with the spherical shape and the elastic stiffness tensor $\mathbb{C}$.

Finally, consider the case of an ellipsoid I. It is interesting to compare [2.56] for a piecewise constant polarization field $\boldsymbol{\tau}^I \chi^I(\underline{z}')$ and [2.45]. This yields the following result:

$$(\forall \underline{z} \in I) \quad \int_I \mathbb{\Gamma}^\infty(\underline{z} - \underline{z}')\, dV_{z'} = \mathbb{P}^I \qquad [2.57]$$

In other words, the function of the left-hand side is uniform in the ellipsoid I, and is equal to the Hill tensor of the latter, as defined by [2.42].

Let us return to the problem [2.49], now with $\mathbf{E}^\infty \neq 0$. We have two loading parameters; the polarization field $\boldsymbol{\tau}$, and the remote strain $\mathbf{E}^\infty$. Due to the linearity, the sought solution $\boldsymbol{\varepsilon}$ is the sum of the contributions of $\boldsymbol{\tau}$ and $\mathbf{E}^\infty$, respectively. In the absence of the polarization field, we observe that the strain field is uniform and equal to the remote strain $\mathbf{E}^\infty$, since the considered medium is homogeneous. The superposition principle therefore readily returns:

$$\boldsymbol{\varepsilon}(\underline{z}) = \mathbf{E}^\infty - \Gamma^\infty * \boldsymbol{\tau}(\underline{z}) \qquad [2.58]$$

2.4.3. *The Green operator of a finite domain*

We now consider the idea of polarization in the case of a finite domain Ω; for which [2.77] is replaced by

$$\begin{array}{lll} \operatorname{div} \boldsymbol{\sigma} = 0 & (\Omega) & (a) \\ \boldsymbol{\sigma} = \mathbb{C} : \boldsymbol{\varepsilon} + \boldsymbol{\tau}(\underline{z}) & (\Omega) & (b) \\ \underline{\xi} = \mathbf{E} \cdot \underline{z} & (\partial\Omega) & (c) \end{array} \qquad [2.59]$$

where $\mathbf{E}$ is a given symmetric second-order tensor. As previously discussed, the elastic stiffness tensor $\mathbb{C}$ is uniform in Ω. The loading is defined by $\mathbf{E}$ and the polarization field $\boldsymbol{\tau}$. The superposition principle states that the strain field $\boldsymbol{\varepsilon}$ solution to [2.59] depends linearly on $\boldsymbol{\tau}$ and $\mathbf{E}$.

In the case $\boldsymbol{\tau} = 0$, it is observed that the uniform strain field $\boldsymbol{\varepsilon} = \mathbf{E}$ is the solution to [2.59]. Therefore, owing to the property of linearity, the strain field in the general case $\boldsymbol{\tau} \neq 0$ takes the form

$$\boldsymbol{\varepsilon}(\underline{z}) = \mathbf{E} - \int_{\Omega} \mathbb{\Gamma}(\underline{z}, \underline{z}') : \boldsymbol{\tau}(\underline{z}')\, dV_{z'} \qquad [2.60]$$

As opposed to $\mathbb{\Gamma}^{\infty}$, the specificity of the finite domain lies in the fact that the kernel $\mathbb{\Gamma}(\underline{z}, \underline{z}')$ is not a function of the vector $\underline{z} - \underline{z}'$. Similarly to [2.56], the Green operator Γ will be defined by:

$$(\Gamma * \boldsymbol{\tau})(\underline{z}) = \int_{\Omega} \mathbb{\Gamma}(\underline{z}, \underline{z}') : \boldsymbol{\tau}(\underline{z}')\, dV_{z'} \qquad [2.61]$$

With this notation, [2.60] takes the same form as [2.58], which is reproduced below for clarity:

$$\boldsymbol{\varepsilon}(\underline{z}) = \mathbf{E} - \Gamma * \boldsymbol{\tau}(\underline{z}) \qquad [2.62]$$

Two useful properties of Γ should be kept in mind. The first interesting property is a consequence of a general result that will be established in section 8.2.1. Considering any displacement field meeting the boundary condition [2.59c], it states that the average[7] strain $\overline{\varepsilon}$ is equal to $\mathbf{E}$ (see equation [8.9]). It then follows from [2.62] that

$$\overline{\Gamma * \boldsymbol{\tau}} = 0 \qquad [2.63]$$

We now observe that no strain is induced by the polarization field if it is a constant. Accordingly, the Green operator is only sensitive to the discrepancy of the local polarization $\boldsymbol{\tau}(\underline{z})$ with respect to the average $\overline{\boldsymbol{\tau}}$:

$$(\Gamma * \boldsymbol{\tau})(\underline{z}) = \int_{\Omega} \mathbb{\Gamma}(\underline{z}, \underline{z}') : \left(\boldsymbol{\tau}(\underline{z}') - \overline{\boldsymbol{\tau}}\right) dV_{\underline{z}'} \qquad [2.64]$$

With regard to the practical use of the Green operator in the case of a finite domain, difficulty arises from the fact that the kernel $\mathbb{\Gamma}(\underline{z}, \underline{z}')$ generally has to be determined numerically. Ponte-Castaneda and Willis propose the approximation of the latter by the kernel of the infinite space $\mathbb{\Gamma}^{\infty}(\underline{z} - \underline{z}')$, in the expression [2.64]:

$$(\Gamma * \boldsymbol{\tau})(\underline{z}) \approx \int_{\Omega} \mathbb{\Gamma}^{\infty}(\underline{z} - \underline{z}') : (\boldsymbol{\tau}(\underline{z}') - \overline{\boldsymbol{\tau}})\, dV_{\underline{z}'} \qquad [2.65]$$

However, note that it certainly would not be correct to replace the kernel $\mathbb{\Gamma}(\underline{z}, \underline{z}')$ by $\mathbb{\Gamma}^{\infty}(\underline{z} - \underline{z}')$ in [2.61]. In fact, considering the case where Ω has an ellipsoidal shape and a uniform polarization $\boldsymbol{\tau}(\underline{z}) = \boldsymbol{\tau}_o$, we would conclude that $\Gamma * \boldsymbol{\tau} = \mathbb{P}^{\Omega} : \boldsymbol{\tau}_o$ as a direct consequence of [2.57]. This result would be in contradiction with [2.64], which yields $\Gamma * \boldsymbol{\tau} = 0$ (since $\boldsymbol{\tau}(\underline{z}') = \overline{\boldsymbol{\tau}}$).

7 The average in the integral sense is defined as in [8.2].

2.4.4. *The inhomogeneity problem*

In the so-called inhomogeneity problem, we again consider an infinite medium comprising a bounded domain I with uniform strain $\mathbf{E}^{\infty}$ at infinity, i.e.:

$$\underline{\xi} \rightarrow \mathbf{E}^{\infty} \cdot \underline{z} \quad \text{for} \quad |\underline{z}| \rightarrow \infty \qquad [2.66]$$

Linear elasticity with no initial stress is assumed. As opposed to the elastic inclusion problem (section 2.4.1), the elastic stiffness now depends on the position vector:

$$\begin{aligned} &\boldsymbol{\sigma} = \mathbb{C}(\underline{z}) : \boldsymbol{\varepsilon} \\ &\underline{z} \in I: \quad \mathbb{C}(\underline{z}) = \mathbb{C}^{I} \\ &\underline{z} \notin I: \quad \mathbb{C}(\underline{z}) = \mathbb{C}_{o} \end{aligned} \qquad [2.67]$$

For solving this new problem, the idea is to see that it can be put in the form [2.34], with

$$\boldsymbol{\tau}^{I}(\underline{z}) = (\mathbb{C}^{I} - \mathbb{C}_{o}) : \boldsymbol{\varepsilon}(\underline{z}) \qquad [2.68]$$

However, as opposed to the inclusion problem in which it is assumed that $\boldsymbol{\tau}^{I}$ is a constant, $(\mathbb{C}^{I} - \mathbb{C}_{o}) : \boldsymbol{\varepsilon}(\underline{z})$ is a priori non-uniform in the domain I since the strain tensor $\boldsymbol{\varepsilon}(\underline{z})$ is expected to depend on $\underline{z}$. Hence, generally the inclusion problem and the inhomogeneity problem are not of the same type. Nevertheless, the case of an ellipsoidal shape (respectively, elliptic shape in plane elasticity) constitutes a counterexample of great interest. This is a direct consequence of the property stated at the end of section 2.4.1, namely that the strain state induced in the domain I by a uniform initial stress is also uniform. Indeed, combining [2.43] and [2.68], we find that[8]

$$\varepsilon^I = \mathbf{E}^\infty - \mathbb{P}^I : (\mathbb{C}^I - \mathbb{C}_o) : \varepsilon^I \qquad [2.69]$$

In conclusion, we shall keep in mind that the strain tensor ε^I induced in an ellipsoidal (respectively, elliptic) inhomogeneity I embedded in an infinite homogeneous medium with uniform strain $\mathbf{E}^\infty$ at infinity reads:

$$\varepsilon^I = (\mathbb{I} + \mathbb{P}^I : (\mathbb{C}^I - \mathbb{C}_o))^{-1} : \mathbf{E}^\infty \qquad [2.70]$$

where $\mathbb{I}$ is the fourth-order unit tensor.

It is helpful for future use to derive the deformation gradient in the domain I. Like strain, it is uniform throughout the domain I. It is readily obtained from [2.68] and [2.47]:

$$\nabla \underline{\xi}^I = -\mathbb{R}^I : (\mathbb{C}^I - \mathbb{C}_o) : \varepsilon^I \qquad [2.71]$$

Introducing [2.70] into the above equation returns:

$$\nabla \underline{\xi}^I = -\mathbb{R}^I : (\mathbb{C}^I - \mathbb{C}_o) : (\mathbb{I} + \mathbb{P}^I : (\mathbb{C}^I - \mathbb{C}_o))^{-1} : \mathbf{E}^\infty \qquad [2.72]$$

which should be compared to [2.47].

2.4.5. *The inhomogeneity problem with stress boundary conditions*

We again consider an infinite medium comprising a bounded domain I. The state equation reads as in [2.67].

8 The Hill tensor $\mathbb{P}^I$ is related to the elastic properties defined by the stiffness tensor $\mathbb{C}_o$, through its definition [2.42] from the Green function of the same material.

However, instead of the displacement boundary conditions [2.66], we now consider the remote stress boundary conditions[9]:

$$\sigma \rightarrow \boldsymbol{\Sigma}^{\infty} \quad \text{for} \quad |\underline{z}| \rightarrow \infty \qquad [2.73]$$

We observe that the stress field in the inhomogeneity problem solved in section 2.4.4 tends toward $\mathbb{C}_o : \mathbf{E}^{\infty}$. It follows that the solution to the new inhomogeneity problem defined by [2.73] is readily obtained by replacing $\mathbf{E}^{\infty}$ by $\mathbb{C}_o^{-1} : \boldsymbol{\Sigma}^{\infty}$. Eventually, in the case of an ellipsoidal (respectively, elliptic) inclusion, the strain tensor ε^I induced in the latter takes the form:

$$\varepsilon^I = (\mathbb{I} + \mathbb{P}^I : (\mathbb{C}^I - \mathbb{C}_o))^{-1} : \mathbb{C}_o^{-1} : \boldsymbol{\Sigma}^{\infty} \qquad [2.74]$$

In turn, the stress $\sigma^I = \mathbb{C}^I : \varepsilon^I$ induced in the domain I is directly obtained by:

$$\sigma^I = (\mathbb{I} + \mathbb{Q}^I : (\mathbb{S}^I - \mathbb{S}_o))^{-1} : \boldsymbol{\Sigma}^{\infty} \qquad [2.75]$$

with

$$\mathbb{Q}^I = \mathbb{C}_o - \mathbb{C}_o : \mathbb{P}^I : \mathbb{C}_o \qquad [2.76]$$

Like $\mathbb{P}^I$, $\mathbb{Q}^I$ is a symmetric tensor.

2.4.6. *The infinite heterogeneous elastic medium*

We again consider an infinite linear elastic medium. But unlike the earlier sections, the elastic properties are neither uniform (as in section 2.4.1) nor piecewise constant (as in

9 Note that these remote conditions deal with the stress tensor and not with the stress vector.

section 2.4.4). In this section, the inhomogeneity is characterized by the field of stiffness tensors $\mathbb{C}(\underline{z})$. When a remote strain $\mathbf{E}^{\infty}$ is applied at infinity, the unknown local stress and strain fields are the solutions to:

$$\begin{array}{ll} \operatorname{div} \boldsymbol{\sigma} = 0 & (a) \\ \boldsymbol{\sigma} = \mathbb{C}(\underline{z}) : \boldsymbol{\varepsilon} & (b) \\ \underline{\xi} = \mathbf{E}^{\infty} \cdot \underline{z} \text{ when } \underline{z} \to \infty & (c) \end{array} \qquad [2.77]$$

The strain field solution to [2.77] can be expressed formally by means of the Green tensor. To do so, let us introduce a so-called reference elastic stiffness tensor $\mathbb{C}_o$ and consider the associated Green operator, denoted by Γ_o and defined as in section 2.4.2. The idea is a generalization of the reasoning presented in section 2.4.4: it is observed that the local state equation can be put in the form:

$$\boldsymbol{\sigma} = \mathbb{C}(\underline{z}) : \boldsymbol{\varepsilon} = \mathbb{C}_o : \boldsymbol{\varepsilon} + \tau(\underline{z}) \quad \text{with} \quad \tau(\underline{z}) = \delta\mathbb{C} : \boldsymbol{\varepsilon} \qquad [2.78]$$

with $\delta\mathbb{C} = \mathbb{C}(\underline{z}) - \mathbb{C}_o$. According to this form, the solution to [2.77] appears to be the solution to [2.49] as well. It therefore takes the form [2.58]:

$$\boldsymbol{\varepsilon} + \Gamma_o * (\delta\mathbb{C} : \boldsymbol{\varepsilon}) = \mathbf{E}^{\infty} \qquad [2.79]$$

This very general equation [DED 73] implicitly characterizes the solution of any problem of heterogeneous elasticity. However, explicit solutions are rare. The case of the ellipsoidal inhomogeneity is a very important example and has been described in section 2.4.4. Using the notations of this section, let the material surrounding the inhomogeneity be used as the reference (stiffness tensor $\mathbb{C}_o$). This implies that $\delta\mathbb{C} = 0$ in [2.79], except in the inhomogeneity I. The idea is to seek a solution to this equation, which should be a constant ε^I in the ellipoidal region I. Recalling [2.57], we can easily see that [2.69] is retrieved.

2.5. Hill tensor for the elliptic inclusion

Considering a cartesien orthonormal frame, the Hill tensor at point $P(z_1, z_2)$ is defined as:

$$P_{ijkl}(\underline{z}) = -\left(\frac{\partial^2}{\partial z_j \partial z_k}\left(\int_I G_{il}(\underline{z} - \underline{u})\, dS_u\right)\right)_{(ij)(kl)} \qquad [2.80]$$

with $dS_u = du_1 du_2$. In the isotropic case, we can show that the 2D Green function is given by (see equation [2.21]):

$$G_{il}(\underline{z} - \underline{u}) = \frac{1}{8\pi\mu(1-\nu)}\,(r_{,i}r_{,l} - \delta_{il}(3 - 4\nu)\log\frac{r}{r_o}) \qquad [2.81]$$

where $r = \sqrt{(z_1 - u_1)^2 + (z_2 - u_2)^2}$ represents the distance MP between point $M(u_1, u_2) \in I$ and point $P(z_1, z_2)$. The purpose of this section is to determine analytical expressions for the Hill tensor when I is an ellipse and $P \in I$.

2.5.1. *Properties of the logarithmic potential*

This section recalls some useful properties of the logarithmic potential related to a domain I. Considering a cartesien orthonormal frame, it is defined at point $P(z_1, z_2)$ as:

$$\varphi(z_1, z_2) = \int_I \log r\, dS \qquad [2.82]$$

where $r = \sqrt{(z_1 - u_1)^2 + (z_2 - u_2)^2}$ represents the distance MP between point $P(z_1, z_2)$ and point $M(u_1, u_2) \in I$, and

$dS = du_1 du_2$. The derivatives of $\varphi(z_1, z_2)$ with respect to z_1 and z_2 therefore read:

$$\frac{\partial \varphi}{\partial z_1}(z_1, z_2) = \int_I \frac{z_1 - u_1}{(z_1 - u_1)^2 + (z_2 - u_2)^2} dS \quad [2.83]$$

and

$$\frac{\partial \varphi}{\partial z_2}(z_1, z_2) = \int_I \frac{z_2 - u_2}{(z_1 - u_1)^2 + (z_2 - u_2)^2} dS \quad [2.84]$$

We now introduce the polar coordinates (r, θ), with point P as a pole. We can easily see that:

$$\cos\theta = \frac{u_1 - z_1}{r}; \quad \sin\theta = \frac{u_2 - z_2}{r} \quad [2.85]$$

so that [2.83] and [2.84] become:

$$\frac{\partial \varphi}{\partial z_1}(z_1, z_2) = -\int_I \frac{\cos\theta}{r} dS; \ \frac{\partial \varphi}{\partial z_2}(z_1, z_2) = -\int_I \frac{\sin\theta}{r} dS \quad [2.86]$$

with $dS = r dr d\theta$.

We now consider the case where I is an elliptic domain:

$$M(u_1, u_2) \in I \Leftrightarrow \left(\frac{u_1}{a}\right)^2 + \left(\frac{u_2}{b}\right)^2 \leq 1 \quad \text{with} \quad a > b \quad [2.87]$$

Let $P(z_1, z_2)$ denote a point located inside this ellipse. We aim to determine the derivatives [2.86] of the logarithmic potential. We look for the equation of the ellipse in the system of polar coordinates with point P as pole in the form $r = \rho(\theta)$. Let $u_1 = z_1 + r\cos\theta$ and $u_2 = z_2 + r\sin\theta$. $\rho(\theta)$ is the positive root of the equation:

$$\left(\frac{z_1 + r\cos\theta}{a}\right)^2 + \left(\frac{z_2 + r\sin\theta}{b}\right)^2 = 1 \quad [2.88]$$

which can be put in the form $Ar^2 + 2Br + C = 0$ with:

$$A = \left(\frac{\cos\theta}{a}\right)^2 + \left(\frac{\sin\theta}{b}\right)^2$$

$$B = \frac{z_1\cos\theta}{a^2} + \frac{z_2\sin\theta}{b^2} \qquad [2.89]$$

$$C = \left(\frac{z_1}{a}\right)^2 + \left(\frac{z_2}{b}\right)^2 - 1$$

Due to the fact that $P \in I$, it follows that $C < 0$, which ensures that $B^2 - AC > 0$ and $\sqrt{B^2 - AC} > B$. Hence, we find that:

$$\rho(\theta) = \frac{-B + \sqrt{B^2 - AC}}{A} > 0 \qquad [2.90]$$

Recalling [2.86], we can see that:

$$\begin{aligned}\frac{\partial\varphi}{\partial z_1}(z_1, z_2) &= \int_0^{2\pi} -\cos\theta\rho(\theta)\, d\theta;\\ \frac{\partial\varphi}{\partial z_2}(z_1, z_2) &= \int_0^{2\pi} -\sin\theta\rho(\theta)\, d\theta\end{aligned} \qquad [2.91]$$

We observe that $\sqrt{B^2 - AC}/A$ is a periodic function of θ with period π. This implies that the corresponding contribution to the integrals of [2.91] vanish. Therefore, the latter simply reads:

$$\frac{\partial\varphi}{\partial z_1}(z_1, z_2) = \int_0^{2\pi} \frac{\dfrac{z_1\cos\theta}{a^2} + \dfrac{z_2\sin\theta}{b^2}}{\left(\dfrac{\cos\theta}{a}\right)^2 + \left(\dfrac{\sin\theta}{b}\right)^2}\cos\theta\, d\theta \qquad [2.92]$$

and

$$\frac{\partial \varphi}{\partial z_2}(z_1, z_2) = \int_0^{2\pi} \frac{\dfrac{z_1 \cos\theta}{a^2} + \dfrac{z_2 \sin\theta}{b^2}}{\left(\dfrac{\cos\theta}{a}\right)^2 + \left(\dfrac{\sin\theta}{b}\right)^2} \sin\theta \, d\theta \qquad [2.93]$$

In the expression of $\partial\varphi/\partial z_1$ (resp. $\partial\varphi/\partial z_2$), it should be observed that the contribution of z_2 (respectively, z_1) vanishes[10]. This proves that $\varphi(z_1, z_2)$ is a quadratic function of the form $\alpha z_1^2 + \beta z_2^2 + \gamma$. After calculation (see section 2.7), we find that:

$$\frac{\partial^2 \varphi}{\partial z_1^2} = 2\alpha = \frac{2\pi X}{1+X}; \; \frac{\partial^2 \varphi}{\partial z_2^2} = 2\beta = \frac{2\pi}{1+X}; \; \frac{\partial^2 \varphi}{\partial z_1 \partial z_2} = 0 \qquad [2.94]$$

with $X = b/a < 1$.

2.5.2. *Integration of the $r_{,i}r_{,l}$ term*

We now address the derivation of the integral

$$R_{\alpha\beta} = \int_I r_{,\alpha} r_{,\beta} \, dS \qquad [2.95]$$

which arises in the expression [2.80] of the components of the Hill tensor $\mathbb{P}$. Recalling the definition of $r = \sqrt{(z_1 - u_1)^2 + (z_2 - u_2)^2}$ and using [2.85], we obtain the following identities:

$$\begin{aligned} r_{,1}r_{,1} &= \cos^2\theta \\ r_{,1}r_{,2} &= \cos\theta \sin\theta \\ r_{,2}r_{,2} &= \sin^2\theta \end{aligned} \qquad [2.96]$$

10 Use the change in variable $\theta' = -\theta$.

that is, $r_{,\alpha}r_{,\beta} = \cos^p\theta\sin^q\theta$, with $p, q \in \{0,1,2\}$. Introducing this result into [2.95] returns:

$$R_{\alpha\beta} = \int_I r_{,\alpha}r_{,\beta}\, r dr d\theta = \frac{1}{2}\int_0^{2\pi} \rho^2(\theta)\cos^p\theta\sin^q\theta\, d\theta \qquad [2.97]$$

with $p+q=2$. Starting from [2.90], we obtain:

$$\rho^2(\theta) = \frac{2B^2 - AC}{A^2} - 2\frac{B\sqrt{B^2-AC}}{A^2} \qquad [2.98]$$

We now take advantage of the fact that $B\sqrt{B^2-AC}/A^2$ takes opposite values in θ and $\theta+\pi$ (see equation [2.89]). This implies that the corresponding contribution in [2.97] vanishes, so that:

$$R_{\alpha\beta} = \frac{1}{2}\int_0^{2\pi} \frac{2B^2 - AC}{A^2}\cos^p\theta\sin^q\theta\, d\theta \qquad [2.99]$$

Observing that:

$$\begin{aligned} 2B^2 - AC = {} & z_1^2\left(\frac{\cos^2\theta}{a^4} - \frac{\sin^2\theta}{a^2b^2}\right) + z_2^2\left(\frac{\sin^2\theta}{b^4} - \frac{\cos^2\theta}{a^2b^2}\right) \\ & +4z_1z_2\frac{\cos\theta\sin\theta}{a^2b^2} + \frac{\cos^2\theta}{a^2} + \frac{\sin^2\theta}{b^2} \end{aligned} \qquad [2.100]$$

we readily see that $R_{\alpha\beta}$ is a quadratic function of $\underline{z}$, the expression of which involves integrals of the form

$$I(m,n) = \int_0^{2\pi} \frac{\cos^m\theta\sin^n\theta}{A^2}\, d\theta \qquad [2.101]$$

The values of the relevant integrals can be found in Appendix (section 2.7). It follows that:

$$\int_I r_{,1}r_{,1}\, dS = \frac{\pi\left(z_1{}^2X^2 - z_2{}^2 + a^2(X^2+X)\right)}{(1+X)^2} \qquad [2.102]$$

$$\int_I r_{,2} r_{,2} \, dS = \frac{\pi \left(-z_1{}^2 X^2 + z_2{}^2 + a^2 (X^2 + X^3)\right)}{(1+X)^2} \qquad [2.103]$$

$$\int_I r_{,1} r_{,2} \, dS = \frac{2\pi X}{(1+X)^2} z_1 z_2 \qquad [2.104]$$

2.5.3. *Components of the Hill tensor*

We now combine the results of sections 2.5.1 and 2.5.2 in order to derive the components of the Hill tensor by application of [2.80]. We recall that the domain I is the ellipse defined by [2.87] and that we only consider the case $\underline{z} \in I$.

These results are summarized in equations [2.94] and [2.102]–[2.104]. They return:

$$\begin{aligned}
P_{1111} &= \frac{X\,(6\,X\mu + 3\,k + 7\,\mu)}{2\,(1+X)^2\,(3\,k + 4\,\mu)\,\mu} \\
&= \frac{(1+\nu)\,((2-4\,\nu)\,X + 3 - 4\,\nu)\,X}{2\,(1-\nu)\,(1+X)^2\,E} \\
P_{1122} &= -\frac{X\,(3\,k+\mu)}{2\,(1+X)^2\,(3\,k+4\,\mu)\,\mu} \\
&= -\frac{X\,(1+\nu)}{2\,(1-\nu)\,(1+X)^2\,E} \\
P_{2222} &= \frac{6\,\mu + X(3k+7\mu)}{2\,(1+X)^2\,(3\,k+4\,\mu)\,\mu} \\
&= \frac{(1+\nu)\,(2-4\,\nu + X(3-4\nu))}{2\,(1-\nu)\,(1+X)^2\,E} \\
P_{1212} &= \frac{6\,X\mu + (3k+4\mu)(1+X^2)}{4\,(1+X)^2\,(3\,k+4\,\mu)\,\mu} \\
&= \frac{(1+\nu)\left((1-\nu)\,(1+X^2) + X\,(1-2\,\nu)\right)}{2\,(1-\nu)\,(1+X)^2\,E}
\end{aligned} \qquad [2.105]$$

For the sake of completeness, we also give coefficients R_{1212} and R_{2112} (see equation [2.48]):

$$R_{1212} = \frac{3k + 4\mu + 3X\mu}{2\mu(3k + 4\mu)(1 + X)^2}$$

$$= \frac{(2(1-\nu) + X(1-2\nu))(1+\nu)}{2(1-\nu)(1+X)^2 E} \qquad [2.106]$$

and

$$R_{2112} = \frac{X(3Xk + 3\mu + 4X\mu)}{2\mu(3k + 4\mu)(1 + X)^2}$$

$$= \frac{X(2X(1-\nu) + 1 - 2\nu)(1+\nu)}{2(1-\nu)(1+X)^2 E} \qquad [2.107]$$

2.6. Hill's tensor for the spheroidal inclusion

Considering a three-dimensional orthonormal frame $Oz_1z_2z_3$, the starting point is, again, the expression of P_{ijkl} at point $\underline{z} = (z_1, z_2, z_3)$ as a function of the second derivatives of the Green's function:

$$P_{ijkl}(\underline{z}) = -\left(\frac{\partial^2 \gamma_{il}}{\partial z_j \partial z_k}\right)_{(ij)(kl)} \quad \text{with } \gamma_{il}(\underline{z}) = \int_I G_{il}(\underline{z} - \underline{z}')\, dV_{z'}$$

[2.108]

with $dV_{z'} = dz'_1 dz'_2 dz'_3$. We focus on the case of isotropic elasticity for which the Green's function is given by [2.33]. First, the integrals γ_{il} which appear in [2.108] are to be derived. They involve the following quantities:

$$\psi(\underline{z}) = \int_I \frac{dV_{z'}}{|\underline{z} - \underline{z}'|}; \quad \chi_{il}(\underline{z}) = \int_I \frac{(z_i - z'_i)(z_l - z'_l)}{|\underline{z} - \underline{z}'|^3} dV_{z'} \qquad [2.109]$$

More precisely:

$$\gamma_{il}(\underline{z}) = \frac{1}{8\pi\mu}\left(\frac{7\mu+3k}{4\mu+3k}\psi(\underline{z})\delta_{il} + \frac{\mu+3k}{4\mu+3k}\chi_{il}(\underline{z})\right) \quad [2.110]$$

We now consider the case where I is a spheroid with its symmetry axis parallel to $\underline{e}_3$:

$$M(z'_1, z'_2, z'_3) \in I \Leftrightarrow \left(\frac{z'_1}{a}\right)^2 + \left(\frac{z'_2}{a}\right)^2 + \left(\frac{z'_3}{c}\right)^2 \leq 1 \quad \text{with} \quad a > c \quad [2.111]$$

the aspect ratio is $X = c/a < 1$ (oblate spheroid). Let $P(z_1, z_2, z_3)$ denote a point located inside this spheroid. Following the same reasoning as in section 2.5.1 (see also [KEL 67]), we seek the equation of the spheroid in the system of spherical coordinates (r, θ, ϕ) with point P as pole[11] in the form $r = \rho(\theta, \phi)$, the angle θ being defined with respect to the direction of $\underline{e}_3$. Let $z'_1 = z_1 + r\sin\theta\cos\phi$, $z'_2 = z_2 + r\sin\theta\sin\phi$ and $z'_3 = z_3 + r\cos\theta$. The quantity $\rho(\theta, \phi)$ is the positive root of the equation:

$$\left(\frac{z_1 + r\sin\theta\cos\phi}{a}\right)^2 + \left(\frac{z_2 + r\sin\theta\sin\phi}{a}\right)^2 + \left(\frac{z_3 + r\cos\theta}{c}\right)^2 = 1 \quad [2.112]$$

which can be put in the form $Ar^2 + 2Br + C = 0$ with:

$$\begin{aligned} A &= \left(\frac{\sin\theta}{a}\right)^2 + \left(\frac{\cos\theta}{c}\right)^2 \\ B &= \frac{z_1\sin\theta\cos\phi}{a^2} + \frac{z_2\sin\theta\sin\phi}{a^2} + \frac{z_3\cos\theta}{c^2} \\ C &= \left(\frac{z_1}{a}\right)^2 + \left(\frac{z_2}{a}\right)^2 + \left(\frac{z_3}{c}\right)^2 - 1 \end{aligned} \quad [2.113]$$

11 $r = \sqrt{(z_1 - z'_1)^2 + (z_2 - z'_2)^2 + (z_3 - z'_3)^2}$.

Owing to the fact that $P \in I$, it follows that $C < 0$, which ensures that $B^2 - AC > 0$ and $\sqrt{B^2 - AC} > B$. Hence, we find that:

$$\rho(\theta, \phi) = \frac{-B + \sqrt{B^2 - AC}}{A} > 0 \qquad [2.114]$$

Recalling that $dV_{z'} = r^2 \sin\theta dr d\theta d\phi$, the integral $\psi(\underline{z})$ of [2.109] can be given in the form:

$$\psi(\underline{z}) = \int_I r \sin\theta d\theta d\phi = \frac{1}{2}\int_0^{\pi} \sin\theta \left(\int_0^{2\pi} \rho^2(\theta, \phi) d\phi\right) d\theta \qquad [2.115]$$

As in [2.98], it follows from [2.114] that:

$$\rho^2(\theta, \phi) = \frac{2B^2 - AC}{A^2} - 2\frac{B\sqrt{B^2 - AC}}{A^2} \qquad [2.116]$$

The change in variable $\theta' = \pi - \theta$, $\phi' = \pi + \phi$ reveals that the contribution to [2.115] of the square root $\sqrt{B^2 - AC}$ in the above expression vanishes. The remaining terms involve integrals listed in Appendix (section 2.8). A similar technique can be applied to the integrals $\chi_{il}(\underline{z})$ of [2.109]:

$$\chi_{il}(\underline{z}) = \frac{1}{2}\int_0^{\pi} \sin\theta d\theta \int_0^{2\pi} \rho^2(\theta, \phi)\lambda_{il}(\theta, \phi) d\phi \qquad [2.117]$$

with

$$\begin{array}{ll} \lambda_{11}(\theta, \phi) = \sin^2\theta \cos^2\phi & \lambda_{23}(\theta, \phi) = \sin\theta \cos\theta \sin\phi \\ \lambda_{22}(\theta, \phi) = \sin^2\theta \sin^2\phi & \lambda_{13}(\theta, \phi) = \sin\theta \cos\theta \cos\phi \\ \lambda_{33}(\theta, \phi) = \cos^2\theta & \lambda_{12}(\theta, \phi) = \sin^2\theta \sin\phi \cos\phi \end{array} \qquad [2.118]$$

2.6.1. *Components of the Hill tensor*

First, the expression of $\gamma_{il,jk}$ (see equation [2.108] and [2.110]) has to be determined. It involves the functions $\psi(\underline{z})$ and $\chi_{il}(\underline{z})$, which are given in Appendix (section 2.8). Then, the symmetrization with respect to $\{i, j\}$ and $\{k, l\}$ is performed:

$$P_{ijkl} = -\frac{1}{4}\left(\gamma_{il,jk} + \gamma_{jl,ik} + \gamma_{ik,jl} + \gamma_{jk,il}\right) \qquad [2.119]$$

It is convenient to introduce P'_{ijkl} defined as:

$$P'_{ijkl} = (\nu - 1)(1 - X^2)^{5/2} P_{ijkl} E \qquad [2.120]$$

The following expressions are obtained:

$$\begin{aligned}
P'_{1111} &= \frac{X\,(1+\nu)}{16}\Big(- \big(4X^2(4\nu - 1) - 16\,\nu + 13\big) \arccos{(X)} \\
&\quad + X\sqrt{1-X^2}\,\big(2X^2(8\nu - 5) - 16\,\nu + 19\big)\Big) \\
P'_{3333} &= \frac{(1+\nu)}{2}\Big(\big(1 - 4\nu + 4X^2(\nu - 1)\big)\, X \arccos(X) \\
&\quad + \sqrt{1-X^2}\,\big(X^2(5 - 4\nu) - 2 + 4\nu\big)\Big) \\
P'_{1122} &= \frac{X(1+\nu)}{16}\big((1 - 4X^2)\arccos(X) + X\sqrt{1-X^2}(1 + 2X^2)\big) \\
P'_{1133} &= -\frac{X(1+\nu)}{4}\big(- (1 + 2X^2)\arccos(X) + 3X\sqrt{1-X^2}\big) \\
P'_{1212} &= -\frac{X(1+\nu)}{16}\Big(+ \arccos(X)\,\big(4X^2(2\nu - 1) + 7 - 8\nu\big) \\
&\quad + X\sqrt{1-X^2}\,\big(8\nu - 9 + 2X^2(3 - 4\nu)\big)\Big) \\
P'_{1313} &= \frac{(1+\nu)}{4}\Big(\arccos(X)\,\big(X^3(1+\nu) + (2 - \nu)X\big) \\
&\quad - \sqrt{1-X^2}\,\big((1 - \nu)(X^4 + 2) + 3X^2\nu\big)\Big)
\end{aligned} \qquad [2.121]$$

Similarly, we introduce

$$R'_{ijkl} = (\nu - 1)(1 - X^2)^{5/2} R_{ijkl} E \qquad [2.122]$$

with

$$R_{ijkl} = -\frac{1}{2}(\gamma_{il,jk} + \gamma_{ik,jl}) \qquad [2.123]$$

The two useful coefficients are:

$$\begin{aligned} R'_{1313} &= \frac{(1+\nu)}{4}\Big(\arccos(X)\left(X(5-4\nu) + (4\nu-2)X^3\right) \\ &\quad + \sqrt{1-X^2}\left(4(\nu-1) + X^2(1-4\nu)\right)\Big) \qquad [2.124] \\ R'_{3113} &= -\frac{X(1+\nu)}{4}\Big(\arccos(X)\left(1 - 2\nu + (2\nu-4)X^2\right) \\ &\quad + \sqrt{1-X^2}\left((1+2\nu)X + 2X^3(1-\nu)\right)\Big) \qquad [2.125] \end{aligned}$$

As expected, it is evident that $2P_{1313} = R_{1313} + R_{3113}$.

2.6.2. *Series expansions of the components of the Hill tensor for flat spheroids*

The following expansions of the P_{ijkl} components with respect to X about $X = 0$ can be derived from [2.121]:

$$P_{1111} = X\frac{(1+\nu)\pi(16\nu - 13)}{32E(\nu - 1)}$$

$$P_{3333} = -\frac{(1+\nu)}{4E(\nu-1)}\left(4(1-2\nu) + \pi X(4\nu - 1)\right)$$

$$P_{1122} = X\frac{\pi(1+\nu)}{32E(\nu-1)}$$

$$P_{1133} = X\frac{\pi(1+\nu)}{8E(\nu-1)}$$

$$P_{1212} = X\frac{\pi(1+\nu)(8\nu-7)}{32E(\nu-1)}$$

$$P_{1313} = \frac{(1+\nu)}{2E}(1-\pi X\frac{2-\nu}{4(1-\nu)}) \quad [2.126]$$

2.7. Appendix

The developments in section 2.5.1 require the determination of integrals of the form:

$$I = \int_0^{2\pi} R(\cos\theta, \sin\theta)\, d\theta \quad [2.127]$$

in which R is a rational function. In view of the application of the theorem of residues, let us introduce the following change in variable:

$$\cos\theta = \frac{1}{2}(z+\frac{1}{z}); \quad \sin\theta = \frac{1}{2i}(z-\frac{1}{z}) \quad [2.128]$$

with $z = e^{i\theta}$ and $dz = izd\theta$. I now reads:

$$I = \oint_{|z|=1} R(\frac{1}{2}(z+\frac{1}{z}), \frac{1}{2i}(z-\frac{1}{z}))\,\frac{dz}{iz} \quad [2.129]$$

Let z_k denote the poles of the function

$$R(\frac{1}{2}(z+1/z), \frac{1}{2i}(z-1/z))/(iz).$$

The theorem of residues implies that:

$$I = 2i\pi \sum_{|z_k|<1} Res(\frac{1}{iz}R(\frac{1}{2}(z+\frac{1}{z}), \frac{1}{2i}(z-\frac{1}{z})), z_k) \quad [2.130]$$

Application of this result returns the following identities:

$$I(2,0) = \int_0^{2\pi} \frac{cos^2\theta}{\left(\frac{\cos\theta}{a}\right)^2 + \left(\frac{\sin\theta}{b}\right)^2} d\theta = \frac{2a^2b}{a+b}\pi \qquad [2.131]$$

$$I(0,2) = \int_0^{2\pi} \frac{sin^2\theta}{\left(\frac{\cos\theta}{a}\right)^2 + \left(\frac{\sin\theta}{b}\right)^2} d\theta = \frac{2ab^2}{a+b}\pi \qquad [2.132]$$

$$I(2,2) = \int_0^{2\pi} \frac{\cos^2\theta\sin^2\theta}{\left(\left(\frac{\cos\theta}{a}\right)^2 + \left(\frac{\sin\theta}{b}\right)^2\right)^2} d\theta = \frac{a^3b^3}{(a+b)^2}\pi \qquad [2.133]$$

$$I(4,0) = \int_0^{2\pi} \frac{\cos^4\theta}{\left(\left(\frac{\cos\theta}{a}\right)^2 + \left(\frac{\sin\theta}{b}\right)^2\right)^2} d\theta = \frac{(2b+a)ba^4}{(a+b)^2}\pi$$

[2.134]

$$I(0,4) = \int_0^{2\pi} \frac{\sin^4\theta}{\left(\left(\frac{\cos\theta}{a}\right)^2 + \left(\frac{\sin\theta}{b}\right)^2\right)^2} d\theta = \frac{(b+2a)b^4a}{(a+b)^2}\pi$$

[2.135]

For illustrative purposes, the derivation of $I(2,0)$ in [2.131] is presented. The change in variables [2.128] leads to:

$$I(2,0) = \oint_{|z|=1} \frac{-i\,(z+i)^2\,(z-i)^2\,b^2a^2}{(b\,z^2+b+a-z^2a)\,(b\,z^2+b+z^2a-a)\,z}\,dz$$

[2.136]

The poles of the rational fraction in the righthand side are:

$$z_o = 0; \quad z_1^{\pm} = \pm\sqrt{\frac{a-b}{a+b}}; \quad z_2^{\pm} = \pm\sqrt{\frac{a+b}{a-b}} \qquad [2.137]$$

Observing that $|z_2^{\pm}| > 1$, [2.131] is obtained from the sum [2.130], which only involves z_o and $z_1^{\pm}$.

2.8. Appendix: derivation of the χ_{ij}

The following integrals can be determined using standard integration techniques.

$$\begin{aligned}
H_1 &= \int_0^{\pi} \frac{\sin^3\theta}{(\dfrac{\sin^2\theta}{a^2} + \dfrac{\cos^2\theta}{c^2})^2} d\theta \\
&= \frac{c^4}{X\sqrt{1-X^2}}\left(\arccos(X)(\frac{1}{X^2} - \frac{1}{1-X^2}) + \frac{1}{X\sqrt{1-X^2}}\right) \\
H_2 &= \int_0^{\pi} \frac{\sin\theta\cos^2\theta}{(\dfrac{\sin^2\theta}{a^2} + \dfrac{\cos^2\theta}{c^2})^2} d\theta \\
&= \frac{c^4}{X(1-X^2)^{3/2}}(\arccos(X) - X\sqrt{1-X^2}) \\
H_3 &= \int_0^{\pi} \frac{\sin\theta}{\dfrac{\sin^2\theta}{a^2} + \dfrac{\cos^2\theta}{c^2}} d\theta \\
&= 2\frac{c^2}{X\sqrt{1-X^2}} \arccos(X)
\end{aligned} \qquad [2.138]$$

$$
\begin{aligned}
H_4 &= \int_0^{\pi} (\frac{\sin^5\theta \sin^2\theta}{a^2} + \frac{\cos^2\theta}{c^2})^2 \, d\theta \\
&= \frac{c^4}{X^3(1-X^2)^{5/2}}((1-4X^2)\arccos(X) \\
&\quad +X\sqrt{1-X^2}(1+2X^2)) \\
H_5 &= \int_0^{\pi} \frac{\sin^3\theta\cos^2\theta}{(\frac{\sin^2\theta}{a^2} + \frac{\cos^2\theta}{c^2})^2} \, d\theta \\
&= \frac{c^4}{X(1-X^2)^{3/2}}(\frac{1+2X^2}{1-X^2}\arccos(X) - \frac{3X}{\sqrt{1-X^2}}) \\
H_6 &= \int_0^{\pi} \frac{\sin^3\theta}{\frac{\sin^2\theta}{a^2} + \frac{\cos^2\theta}{c^2}} \, d\theta \\
&= 2\frac{c^2}{X\sqrt{1-X^2}}(\frac{1}{1-X^2}\arccos(X) - \frac{X}{\sqrt{1-X^2}}) \\
H_7 &= \int_0^{\pi} \frac{\cos^2\theta\sin\theta}{\frac{\sin^2\theta}{a^2} + \frac{\cos^2\theta}{c^2}} \, d\theta \\
&= 2\frac{c^2 X}{(1-X^2)^{3/2}}(\frac{\sqrt{1-X^2}}{X} - \arccos(X)) \\
H_8 &= \int_0^{\pi} \frac{\cos^4\theta\sin\theta}{(\frac{\sin^2\theta}{a^2} + \frac{\cos^2\theta}{c^2})^2} \, d\theta \\
&= 2\frac{c^4 X}{(1-X^2)^{5/2}}(\frac{\sqrt{1-X^2}}{X} - \frac{3}{2}\arccos(X) \\
&\quad +\frac{1}{2}X\sqrt{1-X^2})
\end{aligned}
\qquad [2.139]
$$

These results are used for the derivation of the potential ψ and χ_{ij}:

$$\begin{aligned}
\psi(\underline{z}) &= \int \frac{dV_{z'}}{|\underline{z}-\underline{z}'|} = \pi\big((\frac{H_1}{a^4}-\frac{H_3}{a^2})(z_1^2+z_2^2) \\
&\quad +(2\frac{H_2}{c^4}-\frac{H_3}{c^2})z_3^2+H_3\big) \\
\chi_{11}(\underline{z}) &= \int \frac{(z_1-z_1')^2}{|\underline{z}-\underline{z}'|^3}dV_{z'} = \pi\big(z_1^2(\frac{3H_4}{4a^4}-\frac{H_6}{2a^2}) \\
&\quad +z_2^2(\frac{H_4}{4a^4}-\frac{H_6}{2a^2})+z_3^2(\frac{H_5}{c^4}-\frac{H_6}{2c^2})+\frac{H_6}{2}\big) \\
\chi_{22}(\underline{z}) &= \int \frac{(z_2-z_2')^2}{|\underline{z}-\underline{z}'|^3}dV_{z'} = \pi\big(z_1^2(\frac{H_4}{4a^4}-\frac{H_6}{2a^2}) \\
&\quad +z_2^2(\frac{3H_4}{4a^4}-\frac{H_6}{2a^2})+z_3^2(\frac{H_5}{c^4}-\frac{H_6}{2c^2})+\frac{H_6}{2}\big) \\
\chi_{33}(\underline{z}) &= \int \frac{(z_3-z_3')^2}{|\underline{z}-\underline{z}'|^3}dV_{z'} = \pi\big((z_1^2+z_2^2)(\frac{H_5}{a^4}-\frac{H_7}{a^2}) \\
&\quad +z_3^2(\frac{2H_8}{c^4}-\frac{H_7}{c^2})+H_7\big) \\
\chi_{13}(\underline{z}) &= \int \frac{(z_1-z_1')(z_3-z_3')}{|\underline{z}-\underline{z}'|^3}dV_{z'} = 2\pi z_1 z_3\frac{H_5}{a^2c^2} \\
\chi_{23}(\underline{z}) &= \int \frac{(z_2-z_2')(z_3-z_3')}{|\underline{z}-\underline{z}'|^3}dV_{z'} = 2\pi z_2 z_3\frac{H_5}{a^2c^2} \\
\chi_{12}(\underline{z}) &= \int \frac{(z_1-z_1')(z_2-z_2')}{|\underline{z}-\underline{z}'|^3}dV_{z'} = \pi z_1 z_2\frac{H_4}{2a^4}
\end{aligned} \qquad [2.140]$$

3

Two-dimensional Griffith Crack

In this chapter, we seek the displacement and stress field solutions to the so-called Griffith crack problem in plane strain conditions. The solid is assumed to be linear, elastic and homogeneous.

By definition, the Griffith crack in 2D conditions is a geometrical model in which the crack is described as a segment. The latter is embedded in an infinite plane. More precisely, the crack occupies the segment $z_2 = 0$, $|z_1| < a$ of the plane Oz_1z_2. As usual in plane strain conditions, there is no displacement in the normal direction Oz_3 ($\xi_3 = 0$) and $\partial/\partial z_3 = 0$. Some stress state defined by the tensor[1] $\boldsymbol{\sigma}^\infty$ is applied at infinity.

The presence of the crack is taken into account by the condition $\boldsymbol{\sigma} \cdot \underline{e}_2 = 0$ to be satisfied by the stress field $\boldsymbol{\sigma}$. It expresses that the crack surfaces are free of stress.

1 The stress state $\boldsymbol{\sigma}^\infty$ is assumed to be compatible with plane strain conditions.

Interestingly, if the asymptotic stress state $\boldsymbol{\sigma}^{\infty}$ meets the condition

$$\boldsymbol{\sigma}^{\infty} \cdot \underline{e}_2 = 0 \qquad [3.1]$$

we immediately conclude that the stress field solution $\boldsymbol{\sigma}$ is uniform and equal to $\boldsymbol{\sigma}^{\infty}$. Accordingly, such a loading induces neither a stress or strain singularity, nor displacement jump across the crack. This in turn means that singularities are generated by the "stress vector" $\boldsymbol{\sigma}^{\infty} \cdot \underline{e}_2$, which is defined by its three components σ_{12}^{∞}, σ_{22}^{∞} and σ_{32}^{∞}.

In this chapter, it is assumed that the material surrounding the crack is isotropic. Plane strain conditions therefore prescribe that $\sigma_{32}^{\infty} = 0$. To begin with, the classical proof of the existence of a stress singularity at the crack tip is recalled. This leads to the concept of stress intensity factors K_I in mode I and K_{II} in mode II. The antiplane problem and the stress intensity factor K_{III} will also be briefly introduced. Thereafter, the derivation of K_I and K_{II} is presented. Accordingly, we successively consider a loading defined by $\sigma_{22}^{\infty} \neq 0$ (mode I, section 3.2) and a loading defined by $\sigma_{12}^{\infty} \neq 0$ (mode II, section 3.3).

3.1. Stress singularity at crack tip

This section is devoted to the presentation of concepts of the standard linear elastic fracture mechanics in a two-dimensional context.

We start with the analysis of the stress singularity and the corresponding stress intensity factor-based criterion. Then, we describe the Griffith approach of fracture which introduces the strain energy release concept.

3.1.1. *Stress singularity in plane elasticity: modes I and II*

The existence of a notch in a solid induces a singularity of the stress and strain fields. The singularity at a crack tip is an example of great practical importance. The classical 2D model (see e.g. [LEB 03]) is presented in this section for an isotropic linear elastic solid.

Let us consider a notch in which the two plane edges are free of stress (Figure 3.1). The system of polar coordinates is used, with the origin O located at the intersection of the edges. They respectively correspond to $\theta = +\omega$ and $\theta = -\omega$ (Figure 3.2). The model of the plane crack is obtained when $\omega = \rightarrow \pi^-$. In this case, the two crack lips are in two infinitely close parallel planes defined by the angles $\theta = \pm\pi$.

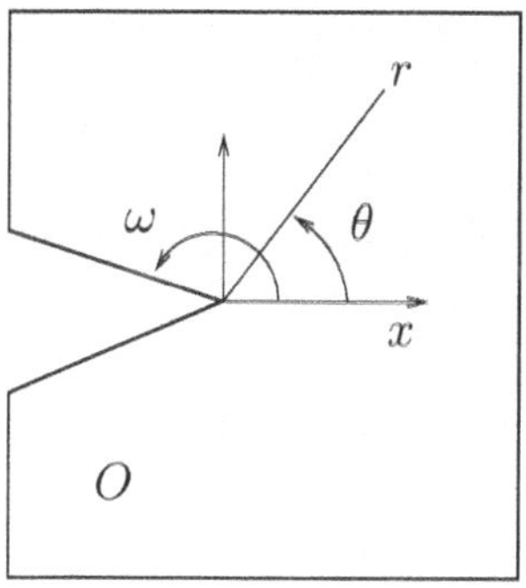

Figure 3.1. *Scheme of a notch and notations*

A variation of the stress components σ_{ij} of the form $r^\alpha g_{ij}(\theta)$ is expected in the neighborhood of O, where α is at this stage an unknown real power. Since the stresses are related to the Airy function by the differential operators of [1.5], the Airy function should read:

$$U(r, \theta) = r^{\alpha+2} f(\theta) \quad [3.2]$$

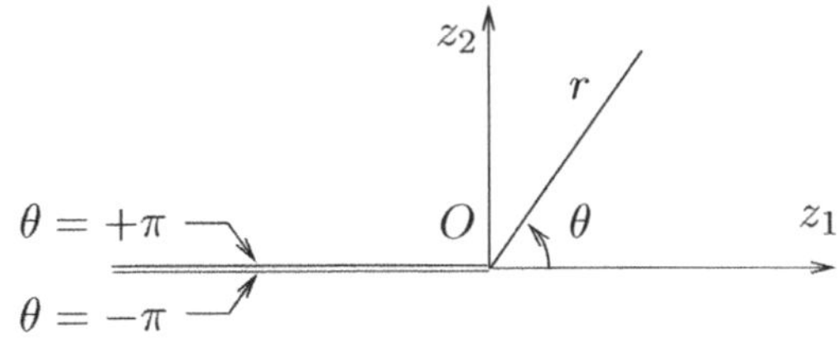

Figure 3.2. *Bidimensional scheme of plane crack*

The existence of a stress singularity amounts to the condition $\alpha < 0$. In addition, the elastic energy stored in the vicinity of the crack is obtained by integration of an energy density which is a quadratic function of the stresses. In the Landau notations, the latter is therefore $O(r^{2\alpha})$. The convergence of the integral in the neighborhood of $r = 0$ (in polar coordinates) requires that $\alpha > -1$. We therefore bear in mind that the power α should belong to the interval $]-1, 0[$.

Two successive applications of [1.3] yield

$$\Delta U = r^{\alpha}\, g(\theta) \quad \text{with} \quad g(\theta) = (\alpha+2)^2 f(\theta) + f''(\theta) \qquad [3.3]$$

and

$$\Delta(\Delta U) = r^{\alpha-2}(\alpha^2 g(\theta) + g''(\theta)) \qquad [3.4]$$

Owing to the biharmonicity condition [1.7], the function $f(\theta)$ of [3.2] is the solution of the differential equation

$$f^{(4)}(\theta) + (\alpha^2 + (\alpha+2)^2) f''(\theta) + \alpha^2(\alpha+2)^2 f(\theta) = 0 \qquad [3.5]$$

Let us consider exponential solutions of [3.5] of the form $a \exp X\theta$. The constant X is the root of the polynom:

$$X^4 + (\alpha^2 + (\alpha+2)^2) X^2 + \alpha^2(\alpha+2)^2 = 0 \qquad [3.6]$$

This implies that $X^2 = -\alpha^2$ or $X^2 = -(\alpha + 2)^2$. The corresponding Airy functions involve sinusoidal functions of θ:

$$\begin{aligned} f(\theta) &= A\cos((\alpha+2)\theta) + B\sin((\alpha+2)\theta) \\ &\quad + C\cos(\alpha\theta) + D\sin(\alpha\theta) \end{aligned} \qquad [3.7]$$

Since the notch edges are free of stress, the coefficients $\sigma_{\theta\theta}$ (normal stress) and $\sigma_{r\theta}$ (tangential stress) are equal to 0 for $\theta = \pm\omega$. Recalling [1.5], these conditions are equivalent to:

$$f'(\theta) = f(\theta) = 0 \quad \text{for} \quad \theta = \pm\omega \qquad [3.8]$$

Observing that

$$\begin{aligned} f'(\theta) &= -A(\alpha+2)\sin((\alpha+2)\theta) + B(\alpha+2)\cos((\alpha+2)\theta) \\ &\quad - C\alpha\sin(\alpha\theta) + D\alpha\cos(\alpha\theta) \end{aligned} \qquad [3.9]$$

the set of conditions [3.8] amounts to the linear system:

$$\begin{pmatrix} \cos(\alpha+2)\omega & \sin(\alpha+2)\omega & \cos\alpha\omega & \sin\alpha\omega \\ \cos(\alpha+2)\omega & -\sin(\alpha+2)\omega & \cos\alpha\omega & -\sin\alpha\omega \\ -(\alpha+2)\sin(\alpha+2)\omega & (\alpha+2)\cos(\alpha+2)\omega & -\alpha\sin\alpha\omega & \alpha\cos\alpha\omega \\ (\alpha+2)\sin(\alpha+2)\omega & (\alpha+2)\cos(\alpha+2)\omega & \alpha\sin\alpha\omega & \alpha\cos\alpha\omega \end{pmatrix} \times \begin{pmatrix} A \\ B \\ C \\ D \end{pmatrix} = \begin{pmatrix} 0 \\ 0 \\ 0 \\ 0 \end{pmatrix} \qquad [3.10]$$

Apart from $A, B, C, D = 0$, the existence of a solution to [3.10] demands that the 4×4 matrix M of the left-hand side be singular, which can be expressed by the condition

$$\begin{aligned} \det M = 0 \Leftrightarrow &((\alpha+1)\sin(2\omega) + \sin(2(\alpha+1)\omega))((\alpha+1)\sin(2\omega) \\ &- \sin(2(\alpha+1)\omega)) = 0 \end{aligned} \qquad [3.11]$$

that is

$$\sin(2(\alpha+1)\omega) = \pm(\alpha+1)\sin(2\omega) \qquad [3.12]$$

Let us now focus on the case of a crack for which $\omega \to \pi^-$. It is readily seen that $\alpha = -1/2$ is the unique solution to [3.12] in the interval $]-1,0[$. Accordingly, the stress field in the vicinity of the crack tip presents a singularity of the type $1/\sqrt{r}$.

For $\alpha = -1/2$, the vectors $(0,-1,0,1)$ and $(1,0,3,0)$ constitute a basis of the kernel of the matrix of [3.10]. The general form of an Airy function yielding a physically relevant singular solution thus reads:

$$\begin{aligned} U(r,\theta) = r^{3/2}\Bigg(A\left(\cos\frac{3\theta}{2} + 3\cos\frac{\theta}{2}\right) \\ +B\left(\sin\frac{3\theta}{2} + \sin\frac{\theta}{2}\right)\Bigg) \end{aligned} \qquad [3.13]$$

The components of $\boldsymbol{\sigma}$ in polar coordinates in the neighborhood of O are derived from [1.5]:

$$\begin{aligned} \sigma_{rr} &= \frac{1}{4\sqrt{2r}}\Bigg(K_I\left(-\cos\frac{3\theta}{2} + 5\cos\frac{\theta}{2}\right) \\ &\quad + K_{II}\left(3\sin\frac{3\theta}{2} - 5\sin\frac{\theta}{2}\right)\Bigg) \\ \sigma_{r\theta} &= \frac{1}{4\sqrt{2r}}\Bigg(K_I\left(\sin\frac{3\theta}{2} + \sin\frac{\theta}{2}\right) \\ &\quad +K_{II}\left(3\cos\frac{3\theta}{2} + \cos\frac{\theta}{2}\right)\Bigg) \\ \sigma_{\theta\theta} &= \frac{1}{4\sqrt{2r}}\Bigg(K_I\left(\cos\frac{3\theta}{2} + 3\cos\frac{\theta}{2}\right) \\ &\quad -3K_{II}\left(\sin\frac{3\theta}{2} + \sin\frac{\theta}{2}\right)\Bigg) \end{aligned} \qquad [3.14]$$

where $K_I = 3A\sqrt{2}$ and $K_{II} = -B\sqrt{2}$. The expression of the local strain tensor is derived from [3.14] in the framework of the assumption of isotropic linear elastic behavior. The strain field exhibits the same singularity of the type $1/\sqrt{r}$ in the neighborhood of point O. The displacement field is obtained by integration (up to a rigid body motion). Since the strain is $O(1/\sqrt{r})$, the displacement is expected to be $O(\sqrt{r})$:

$$\begin{aligned}\xi_r &= \sqrt{\frac{r}{2}}\frac{1}{4\mu}\left(K_I\left(-\cos\frac{3\theta}{2}+(5-8\nu)\cos\frac{\theta}{2}\right)\right.\\&\left.+K_{II}\left(3\sin\frac{3\theta}{2}-(5-8\nu)\sin\frac{\theta}{2}\right)\right)\\ \xi_\theta &= \sqrt{\frac{r}{2}}\frac{1}{4\mu}\left(K_I\left(\sin\frac{3\theta}{2}-(7-8\nu)\sin\frac{\theta}{2}\right)\right.\\&\left.+K_{II}\left(3\cos\frac{3\theta}{2}-(7-8\nu)\cos\frac{\theta}{2}\right)\right)\end{aligned} \quad [3.15]$$

Several comments are worth making:

– the displacement field is discontinuous across the crack. The discontinuity $[\![\underline{\xi}]\!]$ is the relative motion of the upper lip w.r.t. to the lower, that is $\underline{\xi}(r,+\pi)-\underline{\xi}(r,-\pi)$:

$$[\![\underline{\xi}]\!] = \frac{4(1-\nu)}{\mu}\sqrt{\frac{r}{2}}\left(K_{II}\underline{e}_1 + K_I\underline{e}_2\right) \quad [3.16]$$

The coefficients K_I and K_{II} are respectively associated with the normal component of the discontinuity and with the tangential component. These two different loading modes are respectively referred to as mode I and mode II. Mode I is a crack opening loading mode while mode II is a shear loading mode in the crack plane;

– the coefficients K_I and K_{II} cannot be derived from the present analysis the validity of which is restricted to the vicinity of the crack tip (resp. notch). They depend on the geometry of the structure as well as on the remote loading. Analytical expressions of K_I and K_{II} will be derived in sections 3.2 and 3.3 for some simple problems;

– the above expressions of the stresses involve a singularity of the type $1/\sqrt{r}$, which means that the model predicts that the stresses tend to infinity as $r \to 0$ (crack tip)[2]. The existence of infinite stresses is of course not relevant from a physical point of view and has its origin in the geometrical model of a crack as a straight line. In some sense, the elliptic (resp. ellipsoidal) model of crack will correct this shortcoming (see Chapter 4). This result nevertheless suggests that high stress concentrations are expected in the vicinity of geometrical singularities like notches;

– since the stresses are not bounded, the risk of propagation cannot be evaluated by a stress criterion. This is a demarcation line between the theories of brittle vs. ductile failure. Still, the coefficients K_I and K_{II} provide a way to assess the risk of crack propagation. As such, they are referred to as stress intensity factors. The celebrated Irwin criterion $K_I \leq K_I^c$ is written in the form of an upper bound, namely the material toughness K_I^c, prescribed to the stress intensity factor in mode I. A more extensive discussion on crack propagation based on the concept of energy release rate will be proposed in Chapter 7.

3.1.2. *Stress singularity in antiplane problems in elasticity: mode III*

Let us now consider a transformation in which the displacement field is of the form:

$$\underline{\xi} = \xi_3(z_1, z_2)\underline{e}_3 \qquad [3.17]$$

2 Nevertheless, this type of singularity ensures that the energy remains bounded.

We shall refer to this case as an antiplane problem. It follows that the non-vanishing components of the strain tensor are ε_{13} and ε_{23}. Assuming an isotropic linear elastic behavior, the non-vanishing components of the stress tensor are σ_{13} and σ_{23}. In order to meet the momentum balance equation, a stress function $\varphi(z_1, z_2)$ is introduced from which σ_{13} and σ_{23} are derived as follows:

$$\sigma_{13} = -\frac{\partial \varphi}{\partial z_2}(z_1, z_2); \quad \sigma_{23} = \frac{\partial \varphi}{\partial z_1}(z_1, z_2) \tag{3.18}$$

The strain compatibility conditions require that $\varphi(z_1, z_2)$ is a harmonic function:

$$\Delta \varphi(z_1, z_2) = 0 \tag{3.19}$$

Note that, in polar coordinates, the equivalent expression to [3.18] reads:

$$\sigma_{rz} = \frac{\partial \varphi}{\partial r}(r, \theta); \quad \sigma_{\theta z} = \frac{1}{r}\frac{\partial \varphi}{\partial \theta}(r, \theta) \tag{3.20}$$

We now aim to examine the mechanical field in the vicinity of the crack tip in the context of such an antiplane problem. The stress function is sought in polar coordinates. In order to have stress components proportional to r^{α}, this function is taken in the form:

$$\varphi(r, \theta) = r^{\alpha+1} g(\theta) \tag{3.21}$$

From [3.19], it is readily seen that $g(\theta)$ is in the form

$$g(\theta) = A\cos((\alpha + 1)\theta) + B\sin((\alpha + 1)\theta) \tag{3.22}$$

Now, the stress tensor components can be evaluated by reporting [3.21] with [3.22] into [3.20]. The cancellation of the stress component $\sigma_{\theta z}$ on the crack faces $\theta = \pm\pi$ amounts to the condition $g'(\pm\pi) = 0$:

$$\begin{aligned} -A\sin(\alpha+1)\pi + B\cos(\alpha+1)\pi &= 0 \\ A\sin(\alpha+1)\pi + B\cos(\alpha+1)\pi &= 0 \end{aligned} \qquad [3.23]$$

The condition for this linear system to be singular reads:

$$\sin((\alpha+1)\pi)\cos((\alpha+1)\pi) = 0$$

The unique solution $\alpha \in]-1, 0[$ is $\alpha = -\frac{1}{2}$ for which it is readily seen that $A = 0$. The near-tip stress field then reads in antiplane mode (III):

$$\begin{aligned} \sigma_{rz} &= \frac{K_{III}}{\sqrt{2r}}\sin\frac{\theta}{2} \quad (a) \\ \sigma_{rz} &= \frac{K_{III}}{\sqrt{2r}}\cos\frac{\theta}{2} \quad (b) \end{aligned} \qquad [3.24]$$

in which $K_{III} = B/\sqrt{2}$ has been introduced. K_{III} is referred to as the stress intensity factor in this antiplane loading mode.

3.2. Solution to mode I problem

Let us consider a traction in the direction Oz_2 at infinity. It is defined by the stress state $\boldsymbol{\sigma}^\infty$, with[3] $\sigma_{11}^\infty = \sigma_{12}^\infty = 0$ and $\sigma_{22}^\infty = q$ $(q > 0)$. The crack surfaces ($z_2 = 0$ and $|z_1| < a$) are stress free. The corresponding problem is referred to as $\mathcal{P}_I$. Due to the symmetry of the problem with respect to the axis $z_1 = 0$, the displacement field $\underline{\xi}^I(z_1, z_2)$ meets the following conditions:

$$\xi_1^I(-z_1, z_2) = -\xi_1^I(z_1, z_2); \quad \xi_2^I(-z_1, z_2) = \xi_2^I(z_1, z_2) \qquad [3.25]$$

3 Plane strain conditions together with isotropic behavior prescribe that $\sigma_{13} = \sigma_{23} = 0$ and $\sigma_{33} = \nu q$.

Accordingly, only the half space $z_1 > 0$ needs to be considered. The stress boundary conditions defining $\mathcal{P}_I$ read[4]:

$$\mathcal{P}_I : \quad \begin{array}{ll} z_2 = 0 \text{ and } 0 < z_1 < a : & \sigma_{22}^I = \sigma_{12}^I = 0 \\ |\underline{z}| \to \infty : & \sigma_{11}^I, \sigma_{12}^I \to 0 \quad \sigma_{22}^I \to q \end{array} \qquad [3.26]$$

Instead of dealing directly with such boundary conditions, it is convenient to introduce a second problem, referred to as $\mathcal{P}_I'$, defined by $\sigma_{ij} \to 0$ at infinity $(i, j = 1, 2)$ and a uniform pressure q on the crack surfaces:

$$\mathcal{P}_I' : \quad \begin{array}{ll} z_2 = 0 \text{ and } 0 < z_1 < a : & \sigma_{22} = -q \quad ; \quad \sigma_{12} = 0 \\ |\underline{z}| \to \infty : & \sigma_{11}, \sigma_{12}, \sigma_{22} \to 0 \end{array} \qquad [3.27]$$

We finally introduce a third problem $\mathcal{P}_I''$ defined as:

$$\mathcal{P}_I'' : \quad \begin{array}{ll} z_2 = 0 \text{ and } 0 < z_1 < a : & \sigma_{22} = q \quad ; \quad \sigma_{12} = 0 \\ |\underline{z}| \to \infty : & \sigma_{11}, \sigma_{12} \to 0 \quad \sigma_{22} \to q \end{array} \qquad [3.28]$$

Obviously, the stress field solution to problem $\mathcal{P}_I''$ is the uniform stress state $\sigma_{11} = \sigma_{12} = 0$, $\sigma_{22} = q$ in the whole space. It is readily seen that the associated displacement is the following homogeneous transformation:

$$\underline{\xi} = \frac{(1-\nu^2)q}{E} \left(z_2 \underline{e}_2 - \frac{\nu}{1-\nu} z_1 \underline{e}_1 \right) \qquad [3.29]$$

The superposition principle in linear elasticity can be expressed by the formal equation $\mathcal{P}_I = \mathcal{P}_I' + \mathcal{P}_I''$. Accordingly, the stress field solution of $\mathcal{P}_I$ is the sum of the uniform stress field $\sigma_{11} = \sigma_{12} = 0$, $\sigma_{22} = q$ and of the stress field solution of

4 In addition, the symmetry boundary conditions on the $z_1 = 0$ axis prescribe the shear stress $\sigma_{12} = 0$ and the normal component ξ_1 of the displacement: $\xi_1 = 0$.

$\mathcal{P}'_I$. Similarly, the displacement field solution of $\mathcal{P}_I$ is the sum of the solution of $\mathcal{P}'_I$ and of that of $\mathcal{P}''_I$. Solving $\mathcal{P}_I$ thus amounts to solving $\mathcal{P}'_I$.

3.2.1. *Solution of* $\mathcal{P}'_I$

We now seek the solution to $\mathcal{P}'_I$. As just seen before, the solution to $\mathcal{P}_I$ (boundary conditions [3.26]) will be obtained afterward by appropriate superpositions.

The solution of $\mathcal{P}'_I$ (respectively, of $\mathcal{P}_I$) is *a priori* sought in the whole space. Still, symmetry properties make it possible to restrict the study to the half space $z_2 \geq 0$, provided that appropriate boundary conditions be written on the boundary $z_2 = 0$ and $z_1 > a$. In mode I, the latter read:

$$z_2 = 0 \text{ and } z_1 > a : \quad \xi_2 = 0, \, \sigma_{12} = 0 \qquad [3.30]$$

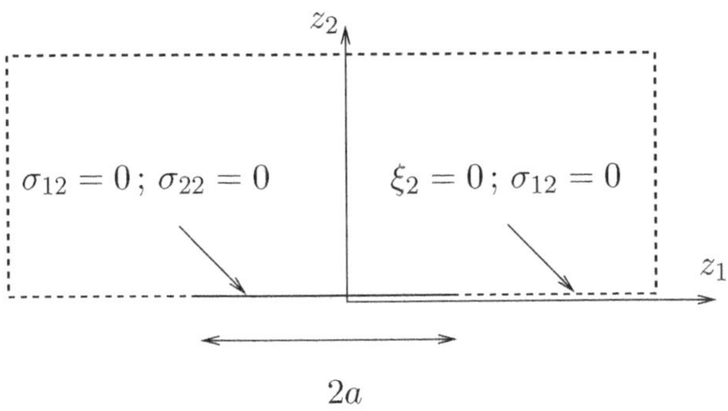

Figure 3.3. *Boundary conditions on the half space boundary in the mode I loading*

In view of the determination of the solution of $\mathcal{P}'_I$ in the half space $z_2 \geq 0$, we consider the displacement potentials described in the Appendix (section 3.5). With $\phi(z_1, z_2)$

denoting a harmonic function, let us first focus on the following displacement field[5]:

$$\underline{\xi}^{(1)} = z_2 \frac{\partial}{\partial z_2} (\underline{\nabla}\, \phi) - (3 - 4\nu)\phi_{,2}\, \underline{e}_2 \qquad [3.31]$$

As already stated, this displacement field satisfies the Navier equation. Its components read:

$$\xi_1^{(1)} = z_2 \phi_{,12}; \quad \xi_2^{(1)} = -(3 - 4\nu)\phi_{,2} + z_2 \phi_{,22} \qquad [3.32]$$

The corresponding strain and stress tensors are defined by:

$$\begin{aligned} \varepsilon_{11}^{(1)} &= z_2 \phi_{,112} \\ \varepsilon_{22}^{(1)} &= z_2 \phi_{,222} - 2(1 - 2\nu)\phi_{,22} \\ \varepsilon_{12}^{(1)} &= z_2 \phi_{,122} - (1 - 2\nu)\phi_{,12} \end{aligned} \qquad [3.33]$$

and

$$\begin{aligned} \sigma_{11}^{(1)} &= 2\mu\,(z_2 \phi_{,112} - 2\nu \phi_{,22}) \\ \sigma_{22}^{(1)} &= 2\mu\,(z_2 \phi_{,222} - 2(1 - \nu)\phi_{,22}) \\ \sigma_{12}^{(1)} &= 2\mu\,(z_2 \phi_{,122} - (1 - 2\nu)\phi_{,12}) \end{aligned} \qquad [3.34]$$

Moving back to section 3.5, we now consider a displacement defined as the gradient of a harmonic function $\psi(z_1, z_2)$ (see [3.114] again):

$$\underline{\xi}^{(2)} = \underline{\nabla}\, \psi \qquad [3.35]$$

---

5 The latter is obtained from [3.114] (see the Appendix, section 3.5) with $\psi = 0$ and $\underline{\Phi} = \phi_{,2}\underline{e}_2$. This form will be justified by forthcoming developments.

associated with the following strains and stresses:

$$\varepsilon_{11}^{(2)} = \psi_{,11} \qquad \varepsilon_{22}^{(2)} = \psi_{,22} \qquad \varepsilon_{12}^{(2)} = \psi_{,12} \qquad [3.36]$$

and

$$\sigma_{11}^{(2)} = 2\mu\psi_{,11} \qquad \sigma_{22}^{(2)} = 2\mu\psi_{,22} \qquad \sigma_{12}^{(2)} = 2\mu\psi_{,12} \qquad [3.37]$$

The displacement field solution to the $\mathcal{P}_I'$ problem is sought in the form of a superposition of $\underline{\xi} = \underline{\xi}^{(1)} + \underline{\xi}^{(2)}$, for an appropriate choice of $\phi(z_1, z_2)$ and $\psi(z_1, z_2)$. In order to meet condition $\sigma_{12} = 0$ on the crack surfaces (see equation [3.27]), we let $\psi = (1 - 2\nu)\phi$. The corresponding components of the displacement vector (ξ_i) and of the stress tensor (σ_{ij}) are:

$$\xi_1 = (1 - 2\nu)\phi_{,1} + z_2\phi_{,12}; \quad \xi_2 = -2(1 - \nu)\phi_{,2} + z_2\phi_{,22} \qquad [3.38]$$

and

$$\begin{aligned} \sigma_{11} &= 2\mu\left(z_2\phi_{,112} - \phi_{,22}\right) \\ \sigma_{22} &= 2\mu\left(z_2\phi_{,222} - \phi_{,22}\right) \\ \sigma_{12} &= 2\mu z_2\phi_{,122} \end{aligned} \qquad [3.39]$$

where the condition $\Delta\phi = 0$ has been used. The condition on the normal stress on the crack surfaces, namely $\sigma_{22} = -q$ (see equation [3.27]) in turn reads:

$$z_2 = 0 \text{ and } 0 < z_1 < a : \quad \sigma_{22} = 2\mu\phi_{,11} = -q \qquad [3.40]$$

In addition, the symmetry of the problem requires that the normal displacement component vanishes, that is $\xi_2 = 0$, in the plane of the crack (see equation [3.30]):

$$z_2 = 0 \text{ and } z_1 > a : \quad \phi_{,2} = 0 \qquad [3.41]$$

The unknown harmonic function $\phi(z_1, z_2)$ solution to [3.40] and [3.41] is now sought in the form of a superposition of elementary harmonic functions[6]:

$$\begin{aligned}\phi(z_1, z_2) &= \int_0^{\infty} \chi(\eta) \exp(-\eta z_2) \cos(\eta z_1)\, d\eta \\ &= \mathcal{F}_c\left(\eta \to \chi(\eta) \exp(-\eta z_2)\right)(z_1)\end{aligned} \quad [3.42]$$

where $\mathcal{F}_c$ (respectively, $\mathcal{F}_s$) denotes the Fourier cosine (respectively, sine) transform, as defined in [3.101]. Solving the problem amounts to the determination of $\chi(\eta)$ meeting [3.40] and [3.41]. It is readily seen that:

$$\begin{aligned}\phi_{,1}(z_1, z_2) &= \mathcal{F}_s\left(\eta \to -\eta\chi(\eta) \exp(-\eta z_2)\right)(z_1) \\ \phi_{,2}(z_1, z_2) &= \mathcal{F}_c\left(\eta \to -\eta\chi(\eta) \exp(-\eta z_2)\right)(z_1) \\ \phi_{,12}(z_1, z_2) &= \mathcal{F}_s\left(\eta \to \eta^2\chi(\eta) \exp(-\eta z_2)\right)(z_1) \\ \phi_{,22}(z_1, z_2) &= \mathcal{F}_c\left(\eta \to \eta^2\chi(\eta) \exp(-\eta z_2)\right)(z_1)\end{aligned} \quad [3.43]$$

Introducing these results into [3.38] yields:

$$\begin{aligned}\xi_1 &= \mathcal{F}_s\left(\eta \to (-(1-2\nu)\eta + z_2\eta^2)\chi(\eta) \exp(-\eta z_2)\right)(z_1) \\ \xi_2 &= \mathcal{F}_c\left(\eta \to (2(1-\nu)\eta + z_2\eta^2)\chi(\eta) \exp(-\eta z_2)\right)(z_1)\end{aligned} \quad [3.44]$$

Introducing $\zeta(\eta) = \eta\chi(\eta)$, we then observe that condition [3.41] combined with [3.43] implies that $\mathcal{F}_c(\zeta)(z_1) = 0$ for $z_1 > a$. In turn, using [3.43], the boundary condition [3.40] reads $2\mu\left(\mathcal{F}_s(\zeta)\right)_{,1}(z_1) = q$. In other words, the boundary conditions require that the unknown function $\zeta(\eta)$ be the solution of the integral equations:

$$\begin{aligned}z_1 > a &: \mathcal{F}_c(\zeta)(z_1) = 0 \quad (a) \\ 0 < z_1 < a &: \left(\mathcal{F}_s(\zeta)\right)_{,1}(z_1) = \frac{q}{2\mu} \quad (b)\end{aligned} \quad [3.45]$$

6 The solution presented hereafter is inspired from [SNE 69].

In order to meet condition [3.45a] automatically, we seek $\zeta(\eta)$ in the form:

$$\zeta(\eta) = \int_0^a J_o(\eta t) f(t)\, dt \qquad [3.46]$$

where J_o is the zero-order Bessel function (see the Appendix, section 3.4, for the definition and properties of J_o). Indeed, considering [3.97], $\mathcal{F}_c(\zeta)(z_1)$ then reads[7]:

$$\int_0^a f(t) \left(\int_0^\infty J_o(\eta t) \cos(\eta z_1)\, d\eta \right) dt = \int_0^a \frac{\mathcal{Y}(t - z_1)}{\sqrt{t^2 - z_1^2}} f(t)\, dt \qquad [3.47]$$

which is clearly equal to 0 for $z_1 > a$. In turn, [3.45b] reads :

$$0 < z_1 < a : \quad \frac{d}{dz_1} \left(\int_0^\infty \left(\int_0^a J_o(\eta t) f(t)\, dt \right) \sin \eta z_1\, d\eta \right) = \frac{q}{2\mu} \qquad [3.48]$$

Recalling [3.98], the integral in the right-hand side of the above equation also takes the form:

$$\int_0^a \left(\int_0^\infty J_o(\eta t) \sin \eta z_1\, d\eta \right) f(t)\, dt = \int_0^a \frac{\mathcal{Y}(z_1 - t)}{\sqrt{z_1^2 - t^2}} f(t)\, dt = \int_0^{z_1} \frac{f(t)\, dt}{\sqrt{z_1^2 - t^2}} \qquad [3.49]$$

7 $\mathcal{Y}(t)$ is the Heaviside function.

where the condition $z_1 < a$ has been used. Eventually, the unknown $f(t)$ in expression [3.46] of $\zeta(\eta)$ is characterized by the integral equation:

$$0 < z_1 < a: \quad \frac{d}{dz_1}\left(\int_0^{z_1} \frac{f(t)\,dt}{\sqrt{z_1^2 - t^2}}\right) = \frac{q}{2\mu} \qquad [3.50]$$

which can be integrated[8]:

$$0 < z_1 < a: \quad \int_0^{z_1} \frac{f(t) - f(0)}{\sqrt{z_1^2 - t^2}}\,dt = \frac{qz_1}{2\mu} \qquad [3.51]$$

In order to solve [3.51] with respect to $f(t)$, let us introduce

$$g(t) = \int_0^t \big(f(u) - f(0)\big)\,du \qquad [3.52]$$

It appears that $g(t)$ is a solution to [3.103] with $p(\rho) = q\rho/(2\mu)$. It follows from [3.110] (Appendix, section 3.4) that:

$$g(z_1) = \frac{qz_1^2}{4\mu}; \quad f(z_1) = \frac{qz_1}{2\mu} + F \qquad [3.53]$$

with $F = f(0)$. It should be observed that [3.50] does not determine $f(t)$ in a unique way since the constant F remains arbitrary. In fact, let ζ_F denote the contribution of the latter to the function ζ defined by [3.46]:

$$\zeta_F(\eta) = F\int_0^a J_o(\eta t)\,dt \qquad [3.54]$$

8 Note that:

$$\lim_{z\to 0}\int_0^z \frac{f(t)\,dt}{\sqrt{z^2 - t^2}} = \frac{\pi}{2}f(0)$$

In turn, the contribution $\underline{\xi}^F$ of ζ_F to the displacement field can be found in replacing $\eta\chi(\eta)$ by the above expression in [3.44]. In particular, for $z_2 = 0$ and $z_1 < a$, it is found that:

$$\frac{\xi_2^F(z_1, 0)}{2(1-\nu)} = \mathcal{F}_c(\zeta_F)(z_1) = F \int_0^a \left(\int_0^\infty J_o(\eta t) \cos \eta z_1 \, d\eta \right) dt \tag{3.55}$$

that is, considering [3.97]

$$\frac{\xi_2^F(z_1, 0)}{2(1-\nu)} = F \int_{z_1}^a \frac{dt}{\sqrt{t^2 - z_1^2}} = F \log \frac{a + \sqrt{a - z_1^2}}{z_1} \tag{3.56}$$

It thus appears that the displacement is infinite for $z_1 = 0$, unless $F = 0$. This determination is therefore adopted. Accordingly, we have from [3.46]:

$$\zeta(\eta) = \eta\chi(\eta) = \frac{q}{2\mu} \int_0^a t J_o(\eta t) \, dt \tag{3.57}$$

which can now be used in [3.44] in order to determine the displacement field solution of problem $\mathcal{P}_I'$ (boundary conditions [3.27]). An analytical expression can be derived on the crack plane ($z_2 = 0$):

$$\begin{aligned} \frac{-2\mu}{(1-2\nu)q} \xi_1 &= \int_0^\infty \left(\int_0^a t J_o(\eta t) dt \right) \sin \eta z_1 \, d\eta \\ &= \int_0^a t \left(\int_0^\infty J_o(\eta t) \sin \eta z_1 \, d\eta \right) dt \\ \frac{\mu}{(1-\nu)q} \xi_2 &= \int_0^\infty \left(\int_0^a t J_o(\eta t) dt \right) \cos \eta z_1 \, d\eta \\ &= \int_0^a t \left(\int_0^\infty J_o(\eta t) \cos \eta z_1 \, d\eta \right) dt \end{aligned} \tag{3.58}$$

Considering [3.97] and [3.98], this first yields the displacement on the upper crack surface ($0 < z_1 < a$, $z_2 = 0$):

$$0 < z_1 < a: \quad \begin{aligned} \xi_1 &= -\frac{q(1-2\nu)}{2\mu}\int_0^{z_1} \frac{t\,dt}{\sqrt{z_1^2 - t^2}} = -\frac{q(1-2\nu)}{2\mu} z_1 \\ \xi_2 &= \frac{q(1-\nu)}{\mu}\int_{z_1}^{a} \frac{t\,dt}{\sqrt{t^2 - z_1^2}} = \frac{q(1-\nu)}{\mu}\sqrt{a^2 - z_1^2} \end{aligned} \qquad [3.59]$$

On the line $z_1 > a$, $z_2 = 0$, we obtain:

$$z_1 > a: \quad \begin{aligned} \xi_1 &= -\frac{q(1-2\nu)}{2\mu}\int_0^{a} \frac{t\,dt}{\sqrt{z_1^2 - t^2}} \\ &= -\frac{q(1-2\nu)}{2\mu}\left(z_1 - \sqrt{z_1^2 - a^2}\right) \\ \xi_2 &= 0 \end{aligned} \qquad [3.60]$$

Similarly, the stress field solution of problem $\mathcal{P}'_I$ can be determined analytically from [3.39] combined with [3.43] in the $z_2 = 0$ plane:

$$\begin{aligned} \sigma_{11} &= -2\mu \mathcal{F}_c\left(\eta \to \eta\zeta(\eta)\right)(z_1) \\ \sigma_{22} &= \sigma_{11} \\ \sigma_{12} &= 0 \end{aligned} \qquad [3.61]$$

that is, recalling [3.46] with $f(t) = qt/2\mu$:

$$\sigma_{11} = \sigma_{22} = -q\int_0^{\infty} \eta \cos \eta z_1 \left(\int_0^a tJ_o(\eta t)\,dt\right) d\eta \qquad [3.62]$$

or

$$\sigma_{11} = \sigma_{22} = -q\frac{d}{dz_1}\left(\int_0^a \left(\int_0^{\infty} \sin \eta z_1 J_o(\eta t)\,d\eta\right) tdt\right) \qquad [3.63]$$

Considering [3.98] and that $z_1 > a$, we eventually obtain:

$$\sigma_{11} = \sigma_{22} = q(\frac{z_1}{\sqrt{z_1^2 - a^2}} - 1); \quad \sigma_{12} = 0 \qquad [3.64]$$

3.2.2. *Solution of* $\mathcal{P}_I$

The displacement field solution of problem $\mathcal{P}_I$ (boundary conditions [3.26]) is denoted by $\underline{\xi}^I$. As already stated, it is obtained from the superposition of the fields of [3.29] and of [3.59] and [3.60]. On the upper crack surface ($z_2 = 0^+$), it is found that:

$$z_2 = 0^+ \;; 0 < z_1 < a : \; \underline{\xi}^I = \frac{q(1-\nu^2)}{E}(-z_1\underline{e}_1 + 2\sqrt{a^2 - z_1^2}\underline{e}_2) \qquad [3.65]$$

In turn, the stress field solution of problem $\mathcal{P}_I$, denoted by $\boldsymbol{\sigma}^I$, is obtained in the plane $z_2 = 0$ from the superposition of the field defined by [3.64] with the uniform field defined by $\sigma_{11} = \sigma_{12} = 0$ and $\sigma_{22} = q$:

$$z_2 = 0 \text{ and } \mathrm{z}_1 > \mathrm{a}: \quad \begin{aligned} \sigma_{11}^I &= q(\frac{z_1}{\sqrt{z_1^2 - a^2}} - 1) \\ \sigma_{22}^I &= q\frac{z_1}{\sqrt{z_1^2 - a^2}} \\ \sigma_{12}^I &= 0 \end{aligned} \qquad [3.66]$$

The above equation reveals the singularity of the normal stress σ_{22}^I in the vicinity of the crack tip $z_1 \rightarrow a^+$. It is convenient to capture this singular behavior by a scalar referred to as stress intensity factor and defined as $K_I = \lim_{z_1 \rightarrow a^+} \sqrt{2(z_1 - a)}\sigma_{22}^I$. It is readily seen from [3.66] that:

$$K_I = q\sqrt{a} \qquad [3.67]$$

3.2.3. *Displacement jump across the crack surfaces*

It is recalled that the above expression [3.65] of the displacement holds in the half space $z_2 \geq 0$. In order to derive the displacement jump $[\![\underline{\xi}^I]\!]$, we need to derive the solution in the lower half space $z_2 \leq 0$. Furthermore, we have to check that the displacement field is continuous across the line $z_2 = 0$, except on the crack surfaces, that is on the segment $z_2 = 0$; $|z_1| < a$.

For clarity, we introduce the notation $\underline{\xi}^{I+}$ (respectively, $\underline{\xi}^{I-}$) for the solution in the half space $z_2 \geq 0$ (respectively, $z_2 \leq 0$). $\underline{\xi}^{I+}$ is the field derived above and given in [3.65]. The symmetry properties of the problem suggest defining $\underline{\xi}^{I-}$ (half space $z_2 < 0$) as:

$$\begin{aligned} \xi_1^{I-}(z_1, z_2) &= \quad \xi_1^{I+}(z_1, -z_2) \\ \xi_2^{I-}(z_1, z_2) &= -\xi_2^{I+}(z_1, -z_2) \end{aligned} \qquad [3.68]$$

It is readily verified that $\underline{\xi}^{I-}$ meets the Navier equation. Accordingly, the displacement jump $[\![\underline{\xi}^I]\!]$ reads:

$$[\![\underline{\xi}^I]\!](z_1) = \underline{\xi}^{I+}(z_1, 0) - \underline{\xi}^{I-}(z_1, 0) = 2\xi_2^{I+}(z_1, 0)\underline{e}_2 \qquad [3.69]$$

We note from [3.29] and [3.60] that $\xi_2^{I+} = 0$ on the line $z_2 = 0$, $z_1 > a$. Hence, [3.69] confirms that there is no displacement discontinuity on this line. In turn, the displacement jump[9] across the crack surfaces ($|z_1| < a$) is derived from [3.65]:

$$|z_1| < a : \quad [\![\underline{\xi}^I]\!] = 4(1 - \nu^2)\frac{\sigma_{22}^{\infty}}{E}\sqrt{a^2 - z_1^2}\,\underline{e}_2 \qquad [3.70]$$

9 It is recalled that, in this section, q stands for σ_{22}^{∞}.

Owing to [3.67], this result is fully consistent with [3.16] since, for $z_1 \to a$, $\sqrt{a^2 - z_1^2} \approx \sqrt{2a}\sqrt{a - z_1}$ $(r = a - z_1)$.

3.3. Solution to mode II problem

Let us now consider a deviatoric stress state at infinity, with $\sigma_{11}^\infty = \sigma_{22}^\infty = 0$ and $\sigma_{12}^\infty = q$ $(q > 0)$. Again, the crack surfaces are stress free. The corresponding problem is referred to as $\mathcal{P}_{II}$. The stress field solution is denoted by $\sigma^{II}(z_1, z_2)$. The displacement field solution denoted by $\underline{\xi}^{II}(z_1, z_2)$ is expected to meet the following conditions:

$$\xi_1^I(-z_1, z_2) = \xi_1^I(z_1, z_2) \quad ; \quad \xi_2^I(-z_1, z_2) = -\xi_2^I(z_1, z_2) \qquad [3.71]$$

In linear isotropic elasticity, it is therefore possible to restrict the study to the domain $z_1 \geq 0$ with the conditions $\xi_2^{II} = 0$ and $\sigma_{11}^{II} = 0$ on the axis $z_1 = 0$. The stress boundary conditions in problem $\mathcal{P}_{II}$ read:

$$\mathcal{P}_{II}: \begin{array}{ll} z_2 = 0 \text{ and } 0 < z_1 < a: & \sigma_{22}^{II} = \sigma_{12}^{II} = 0 \\ |\underline{z}| \to \infty: & \sigma_{11}^{II}, \sigma_{22}^{II} \to 0 \quad \sigma_{12}^{II} \to q \end{array} \qquad [3.72]$$

By analogy with section 3.2, it is convenient to introduce a second problem, referred to as $\mathcal{P}'_{II}$, defined by $\sigma_{ij} \to 0$ at infinity $(i, j = 1, 2)$ and a uniform shear stress $-q$ on the crack surfaces:

$$\mathcal{P}'_{II}: \begin{array}{ll} z_2 = 0 \text{ and } 0 < z_1 < a: & \sigma_{12} = -q; \quad \sigma_{22} = 0 \\ |\underline{z}| \to \infty: & \sigma_{11}, \sigma_{12}, \sigma_{22} \to 0 \end{array} \qquad [3.73]$$

We also introduce problem $\mathcal{P}''_{II}$ defined by:

$$\mathcal{P}''_{II}: \begin{array}{ll} z_2 = 0 \text{ and } 0 < z_1 < a: & \sigma_{12} = q; \quad \sigma_{22} = 0 \\ |\underline{z}| \to \infty: & \sigma_{11}, \sigma_{22} \to 0 \quad \sigma_{12} \to q \end{array} \qquad [3.74]$$

The stress field solution to $\mathcal{P}''_{II}$ is the uniform stress state $\sigma_{11} = \sigma_{22} = 0$, $\sigma_{12} = q$ in the whole space, while the corresponding displacement field is the following homogeneous transformation:

$$\underline{\xi} = \frac{q}{\mu} z_2 \underline{e}_1 \qquad [3.75]$$

The superposition principle in linear elasticity can be expressed by the formal equation $\mathcal{P}_{II} = \mathcal{P}''_{II} + \mathcal{P}''_{II}$. Accordingly, the stress field solution of $\mathcal{P}_{II}$ is the sum of the uniform stress field $\sigma_{11} = \sigma_{22} = 0$, $\sigma_{12} = q$ and of the stress field solution of $\mathcal{P}'_{II}$. Similarly, the displacement field solution of $\mathcal{P}_{II}$ is the sum of the solution of $\mathcal{P}'_{II}$ and of that of $\mathcal{P}''_{II}$. Solving $\mathcal{P}_{II}$ therefore amounts to solving $\mathcal{P}'_{II}$.

3.3.1. *Solution of* $\mathcal{P}'_{II}$

We now seek the solution to $\mathcal{P}'_{II}$. The methodology is strictly identical to that used in section 3.2. The study is restricted to the half space $z_2 \geq 0$. Therefore, additional boundary conditions are introduced on the boundary $z_2 = 0$ and $z_1 > a$:

$$z_2 = 0 \text{ and } z_1 > a : \quad \xi_1 = \sigma_{22} = 0 \qquad [3.76]$$

We again combine displacement fields $\underline{\xi}^{(1)}$ and $\underline{\xi}^{(2)}$ of the form [3.31] and [3.35], $\phi(z_1, z_2)$ and $\psi(z_1, z_2)$ denoting harmonic functions: The displacement field solution of $\mathcal{P}'_{II}$ is sought as a superposition of $\underline{\xi} = \underline{\xi}^{(1)} + \underline{\xi}^{(2)}$, for an appropriate choice of $\phi(z_1, z_2)$ and $\psi(z_1, z_2)$. In order to meet condition $\sigma_{22} = 0$ on the crack surfaces (see equation [3.73]), it is readily seen from [3.34] and [3.37] that we have to choose $\psi = 2(1-\nu)\phi$. The corresponding displacement and stress components are:

$$\xi_1 = 2(1-\nu)\phi_{,1} + z_2\phi_{,12}; \quad \xi_2 = -(1-2\nu)\phi_{,2} + z_2\phi_{,22} \qquad [3.77]$$

which replaces [3.38] and

$$\begin{aligned} \sigma_{11} &= 2\mu\left(z_2\phi_{,112} - 2\phi_{,22}\right) \\ \sigma_{22} &= 2\mu z_2\phi_{,222} \\ \sigma_{12} &= 2\mu\left(z_2\phi_{,122} + \phi_{,12}\right) \end{aligned} \quad [3.78]$$

which replaces [3.39]. The condition on the shear stress on the crack surfaces, namely $\sigma_{12} = -q$, reads:

$$z_2 = 0 \text{ and } 0 < z_1 < a : \quad \sigma_{12} = 2\mu\phi_{,12} = -q \quad [3.79]$$

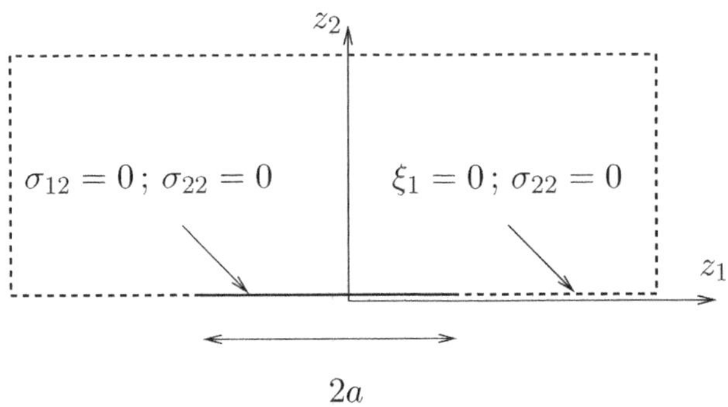

Figure 3.4. *Boundary conditions on the half space boundary in the mode II loading*

Furthermore, the half space boundary condition [3.76] requires that the displacement component ξ_1 vanishes in the plane of the crack:

$$z_2 = 0 \text{ and } z_1 > a : \quad \phi_{,1} = 0 \quad [3.80]$$

The unknown $\phi(z_1, z_2)$ solution to [3.79] and [3.80] is now sought in the form of a superposition of elementary harmonic functions:

$$\phi(z_1, z_2) = \mathcal{F}_s\left(\eta \to \chi(\eta)\exp(-\eta z_2)\right)(z_1) \quad [3.81]$$

so that [3.43] is replaced by:

$$\begin{aligned}
\phi_{,1}(z_1, z_2) &= \mathcal{F}_c\left(\eta \to \eta\chi(\eta)\exp(-\eta z_2)\right)(z_1) \\
\phi_{,2}(z_1, z_2) &= \mathcal{F}_s\left(\eta \to -\eta\chi(\eta)\exp(-\eta z_2)\right)(z_1) \\
\phi_{,12}(z_1, z_2) &= \mathcal{F}_c\left(\eta \to -\eta^2\chi(\eta)\exp(-\eta z_2)\right)(z_1) \\
\phi_{,22}(z_1, z_2) &= \mathcal{F}_s\left(\eta \to \eta^2\chi(\eta)\exp(-\eta z_2)\right)(z_1)
\end{aligned} \qquad [3.82]$$

Introducing these results into [3.77] yields:

$$\begin{aligned}
\xi_1 &= \mathcal{F}_c\left(\eta \to (2(1-\nu) - z_2\eta)\zeta(\eta)\exp(-\eta z_2)\right)(z_1) \\
\xi_2 &= \mathcal{F}_s\left(\eta \to ((1-2\nu) + z_2\eta)\zeta(\eta)\exp(-\eta z_2)\right)(z_1)
\end{aligned} \qquad [3.83]$$

with $\zeta(\eta) = \eta\chi(\eta)$. We then observe that condition [3.80] combined with [3.82] implies that $\mathcal{F}_c(\zeta)(z_1) = 0$ for $z_1 > a$. In turn, using [3.82], the boundary condition on the shear stress [3.79] reads $2\mu\left(\mathcal{F}_s(\zeta)\right)_{,1}(z_1) = q$. This amounts to saying that the unknown function $\zeta(\eta)$ has to meet the system of integral equations [3.45] encountered previously in the mode I loading whose solution reads:

$$\zeta(\eta) = \frac{q}{2\mu}\int_0^a tJ_o(\eta t)\,dt \qquad [3.84]$$

We now introduce this solution into [3.83] in order to determine the displacement field solution of problem $\mathcal{P}'_{II}$ (boundary conditions [3.73]). An analytical expression can be derived on the crack plane ($z_2 = 0$):

$$\begin{aligned}
\xi_1(z_1, 0) &= 2(1-\nu)\mathcal{F}_c(\zeta)(z_1) \\
\xi_2(z_1, 0) &= (1-2\nu)\mathcal{F}_s(\zeta)(z_1)
\end{aligned} \qquad [3.85]$$

that is,

$$\begin{aligned}
\frac{E}{2q(1-\nu^2)}\xi_1 &= \textstyle\int_0^a t\left(\int_0^\infty J_o(\eta t)\cos\eta z_1\,d\eta\right)dt \\
\frac{E}{q(1+\nu)(1-2\nu)}\xi_2 &= \textstyle\int_0^a t\left(\int_0^\infty J_o(\eta t)\sin\eta z_1\,d\eta\right)dt
\end{aligned} \qquad [3.86]$$

The integrals in the right-hand sides have already been encountered in [3.58]: the displacement of the upper crack surface is found to be:

$$z_2 = 0,\ 0 < z_1 < a: \quad \begin{aligned} \xi_1 &= \frac{2q(1-\nu^2)}{E}\sqrt{a^2 - z_1^2} \\ \xi_2 &= \frac{q(1+\nu)(1-2\nu)}{E} z_1 \end{aligned} \qquad [3.87]$$

In turn, on the line $z_1 > a$, $z_2 = 0$, we obtain:

$$z_1 > a: \quad \begin{aligned} \xi_1 &= 0 \\ \xi_2 &= \frac{q(1+\nu)(1-2\nu)}{E}\left(z_1 - \sqrt{z_1^2 - a^2}\right) \end{aligned} \qquad [3.88]$$

Similarly, the stress field solution of problem $\mathcal{P}'_{II}$ can be determined analytically from [3.78] in the $z_2 = 0$ plane:

$$\begin{aligned} \sigma_{11} &= -4\mu \mathcal{F}_s\left(\eta \to \eta\zeta(\eta)\right)(z_1) \\ \sigma_{22} &= 0 \\ \sigma_{12} &= -2\mu \mathcal{F}_c\left(\eta \to \eta\zeta(\eta)\right)(z_1) \end{aligned} \qquad [3.89]$$

It follows that σ_{12} in [3.89] is identical to the coefficient σ_{11} in [3.61]. Furthermore, it appears that $\sigma_{11} = 4\mu\big(\mathcal{F}_c(\zeta)\big)_{,1}$, which is equal to 0 for $z_1 > a$ (see equation [3.45]):

$$z_1 > a: \quad \sigma_{11} = \sigma_{22} = 0; \quad \sigma_{12} = q\left(\frac{z_1}{\sqrt{z_1^2 - a^2}} - 1\right) \qquad [3.90]$$

3.3.2. *Solution of* $\mathcal{P}_{II}$

The displacement field solution of problem $\mathcal{P}_{II}$ (boundary conditions [3.72]), denoted by $\underline{\xi}^{II}$, is obtained from the superposition of the fields of [3.75] and of [3.87] and [3.88].

On the upper crack surface, we obtain:

$$z_2 = 0^+ ; 0 < z_1 < a : \quad \underline{\xi}^{II} = \frac{q(1-\nu^2)}{E}(2\sqrt{a^2 - z_1^2}\underline{e}_1 + \frac{1-2\nu}{1-\nu} z_1 \underline{e}_2) \qquad [3.91]$$

In turn, the stress field solution of problem $\mathcal{P}_{II}$, denoted by $\boldsymbol{\sigma}^{II}$, is obtained in the plane $z_2 = 0$ from the superposition of the field defined by [3.90] with the uniform field defined by $\sigma_{11} = \sigma_{22} = 0$ and $\sigma_{12} = q$:

$$z_2 = 0 \text{ and } z_1 > a : \quad \begin{array}{l} \sigma_{11}^{II} = \sigma_{22}^{II} = 0 \\ \sigma_{12}^{II} = q \dfrac{z_1}{\sqrt{z_1^2 - a^2}} \end{array} \qquad [3.92]$$

The above equation reveals the singularity of the shear stress σ_{12}^{II} in the vicinity of the crack tip $z_1 \to a^+$. As done in mode I, this singular behavior is captured by a scalar referred to as stress intensity factor and defined as $K_{II} = \lim_{z_1 \to a^+} \sqrt{2(z_1 - a)}\sigma_{12}^{II}$. It is readily seen from [3.66] that:

$$K_{II} = q\sqrt{a} \qquad [3.93]$$

3.3.3. *Displacement jump across the crack surfaces*

Expression [3.91] of the displacement holds in the half space $z_2 \geq 0$. In order to derive the displacement jump $[\![\underline{\xi}^{II}]\!]$, we need to derive the solution in the lower half space $z_2 \leq 0$. We also have to check that the displacement field is continuous across the line $z_2 = 0$, except on the crack surfaces, that is on the segment $z_2 = 0$; $|z_1| < a$.

For clarity, we introduce the notation $\underline{\xi}^{II+}$ (respectively, $\underline{\xi}^{II-}$) for the solution in the half space $z_2 \geq 0$ (respectively,

$z_2 \leq 0$). $\underline{\xi}^{II+}$ is the field derived above and given in [3.91]. The symmetry properties of the problem suggest defining $\underline{\xi}^{II-}$ (half space $z_2 < 0$) from $\underline{\xi}^{II+}$ (half space $z_2 > 0$) as:

$$\begin{aligned} \xi_1^{II-}(z_1, z_2) &= -\xi_1^{II+}(z_1, -z_2) \\ \xi_2^{II-}(z_1, z_2) &= \quad \xi_2^{II+}(z_1, -z_2) \end{aligned} \qquad [3.94]$$

Again, it is readily verified that $\underline{\xi}^{II-}$ meets the Navier equation. Accordingly, the displacement jump $[\![\underline{\xi}^{II}]\!]$ across the line $z_2 = 0$ reads:

$$[\![\underline{\xi}^{II}]\!](z_1) = \underline{\xi}^{II+}(z_1, 0) - \underline{\xi}^{II-}(z_1, 0) = 2\xi_1^{II+}(z_1, 0)\underline{e}_1 \qquad [3.95]$$

Equations [3.75] and [3.88] confirm that there is no displacement discontinuity on the line $z_2 = 0$ in the domain $z_1 > a$. In turn, the displacement jump across the crack surfaces ($|z_1| < a$) is derived from [3.91]:

$$|z_1| < a: \quad [\![\underline{\xi}^{II}]\!] = 4(1 - \nu^2)\frac{\sigma_{12}^{\infty}}{E}\sqrt{a^2 - z_1^2}\underline{e}_1 \qquad [3.96]$$

3.4. Appendix: Abel's integral equation

The following developments take advantage of some classical properties of the zero-order Bessel function $J_o(x)$, which is a solution of Bessel's differential equation $x^2y''(x) + xy'(x) + x^2y = 0$. Some properties of the Hankel transform are also reviewed. More precisely, $\mathcal{Y}$ denoting the Heaviside function, it is recalled that:

$$\int_0^{\infty} J_o(\eta\rho) \cos \eta t \, d\eta = \frac{\mathcal{Y}(\rho - t)}{\sqrt{\rho^2 - t^2}} \qquad [3.97]$$

and

$$\int_0^\infty J_o(\eta\rho)\sin\eta t\,d\eta = \frac{\mathcal{Y}(t-\rho)}{\sqrt{t^2-\rho^2}} \tag{3.98}$$

The Hankel transform of $\phi(x)$, denoted by $\mathcal{H}\phi(\eta)$, is defined as:

$$\mathcal{H}\phi(\eta) = \int_0^\infty x\phi(x)J_o(\eta x)\,dx \tag{3.99}$$

The inverse Hankel transform then reads:

$$\phi(x) = \int_0^\infty \eta\mathcal{H}\phi(\eta)J_o(\eta x)\,d\eta \tag{3.100}$$

The current chapter makes use of the Fourier transforms of $g(\rho)$ defined as:

$$\mathcal{F}_s g(x) = \int_0^\infty g(\rho)\sin\rho x\,d\rho; \quad \mathcal{F}_c g(x) = \int_0^\infty g(\rho)\cos\rho x\,d\rho \tag{3.101}$$

with the inverse transforms:

$$\begin{aligned} g(\rho) &= \frac{2}{\pi}\int_0^\infty \mathcal{F}_s g(x)\sin\rho x\,dx; \\ g(\rho) &= \frac{2}{\pi}\int_0^\infty g(x)\mathcal{F}_c\cos\rho x\,dx \end{aligned} \tag{3.102}$$

Let us now introduce Abel's integral equation in the following form:

$$0<\rho<a: \quad \int_0^\rho \frac{g'(t)}{\sqrt{\rho^2-t^2}}\,dt = p(\rho) \quad \text{with} \quad g(0)=0 \tag{3.103}$$

where $g(t)$ is the unknown function, while $p(\rho)$ is given. Only the values of $g(t)$ for $t \in [0, a]$ are of interest. This amounts to saying that $g(t) = 0$ for $t \notin [0, a]$. Let us recall the following integrals:

$$\int_0^\rho \frac{dt}{\sqrt{\rho^2 - t^2}} = \frac{\pi}{2}; \quad \int_0^\rho \frac{tdt}{\sqrt{\rho^2 - t^2}} = \rho \qquad [3.104]$$

When $p(\rho)$ is a constant P, [3.104] shows that the solution of [3.103] is $g'(t) = 2P/\pi$, that is $g(t) = 2Pt/\pi$. In addition, if $p(\rho) = \rho$, we obtain $g'(t) = t$, that is $g(t) = t^2/2$.

We now replace $\mathcal{Y}(\rho - t)/\sqrt{\rho^2 - t^2}$ in the left-hand side of [3.103] by its expression [3.97]:

$$\begin{aligned} p(\rho) &= \int_0^a g'(t) \left(\int_0^\infty J_o(x\rho) \cos xt \, dx \right) dt \\ &= \int_0^\infty J_o(x\rho) \left(\int_0^a g'(t) \cos xt \, dt \right) dx \end{aligned} \qquad [3.105]$$

An integration by parts in the right-hand side of [3.105] yields, for $0 < \rho < a$:

$$\begin{aligned} p(\rho) = &\int_0^\infty xJ_o(x\rho) \left(\int_0^a g(t) \sin xt \, dt \right) dx \\ &+ \int_0^\infty J_o(x\rho)[g(t) \cos xt]_0^a \, dx \end{aligned} \qquad [3.106]$$

Using the condition $g(0) = 0$ together with [3.97] (recall that $\rho < a$), it appears that the second integral of the right-hand side vanishes:

$$p(\rho) = \int_0^\infty xJ_o(x\rho) \left(\int_0^a g(t) \sin xt \, dt \right) dx \qquad [3.107]$$

This reveals that the solution $g(\rho)$ with $g(0) = 0$ of [3.103] and [3.107] is identical.

$p(\rho)$ is the Hankel transform of $\mathcal{F}_s g(x)$. By application of [3.100], the latter is the Hankel transform of $p(\rho)$:

$$\mathcal{F}_s g(x) = \int_0^{\infty} \rho J_o(\rho x) p(\rho)\, d\rho \qquad [3.108]$$

The last step consists of deriving $g(\rho)$ from [3.108] by means of an inverse Fourier transform (see equation [3.102]):

$$\begin{aligned} g(\rho) &= \frac{2}{\pi} \int_0^{\infty} \sin \rho x \left(\int_0^{\infty} s J_o(sx) p(s)\, ds \right) dx \\ &= \frac{2}{\pi} \int_0^{\infty} p(s) s \left(\int_0^{\infty} J_o(sx) \sin \rho x\, dx \right) ds \end{aligned} \qquad [3.109]$$

Using [3.98] in the right-hand side of the above integral, we eventually obtain the solution $g(\rho)$ to [3.103] as a function of $p(s)$:

$$g(\rho) = \frac{2}{\pi} \int_0^{\rho} \frac{p(s) s}{\sqrt{\rho^2 - s^2}}\, ds \qquad [3.110]$$

3.5. Appendix: Neuber–Papkovitch displacement potentials

Consider an isotropic linear elastic solid (in the 3D framework). The general form of the solutions of the Navier equation (Lamé coefficients λ and μ, Poisson coefficient ν, volume density of forces $\underline{f}$):

$$(\lambda + \mu) \underline{\nabla}(\underline{\nabla} \cdot \underline{\xi}) + \mu \Delta \underline{\xi} + \underline{f} = 0$$

can be expressed as functions of a scalar potential ψ and a vector potential $\underline{\Phi}$. The displacement field is related to ψ and $\underline{\Phi}$ by:

$$\underline{\xi} = \frac{1}{\mu}\left(-\underline{\Phi} + \frac{1}{4(1-\nu)}\underline{\nabla}\left(\underline{z}\cdot\underline{\Phi} + \psi\right)\right) \qquad [3.111]$$

The potentials Φ and $\underline{\psi}$ are solutions to the following field equations:

$$\nabla\cdot(\nabla\underline{\Phi}) = \underline{f}; \quad \Delta\psi = -\underline{z}\cdot\underline{f} \qquad [3.112]$$

In particular, in the absence of volume forces ($\underline{f} = 0$), ψ and the components of Ψ (in Cartesian coordinates) should be harmonic functions:

$$\nabla\cdot(\nabla\underline{\Phi}) = 0; \quad \Delta\psi = 0 \qquad [3.113]$$

In this case, it is of course equivalent to write $\underline{\xi}$ in the form:

$$\underline{\xi} = -4(1-\nu)\underline{\Phi} + \underline{\nabla}\left(\underline{z}\cdot\underline{\Phi} + \psi\right) \qquad [3.114]$$

The 2D framework of plane strains (parallel to Oz_1z_2) is readily obtained with conditions $\partial/\partial z_3 = 0$ and $\Phi_3 = 0$.

4

The Elliptic Crack Model in Plane Strains

An alternative to the 2D Griffith crack model consists of describing the crack of the initial configuration as a flat elliptic cylinder parallel to the Oz_3 axis. The cross-section by a $z_3 = \text{Const.}$ plane is an ellipse the equation of which reads:

$$\frac{z_1^2}{a^2} + \frac{z_2^2}{b^2} = 1 \qquad [4.1]$$

where $X = b/a \ll 1$ is referred to as the aspect ratio. As in Chapter 3, plane strain conditions are assumed.

First, the elliptic geometry is considered with the tools of plane elasticity and conformal mapping (sections 1.7.1 and 1.7.3). This approach is presented in section 4.1. As an alternative, the same problem is considered in section 4.3 within the framework of the Eshelby approach (section 2.4).

4.1. The infinite plane with elliptic hole

For illustrative purposes, section 1.6.2 presented the mechanical response of an infinite domain with a circular hole. In view of application to crack modeling, we now more generally consider the elliptic hole. The notations of Chapter 1 are adopted. In particular, the notations z_1 and z_2 are replaced by x and y, respectively.

We are interested in the case where the domain S (notation of section 1.7.1) is the infinite plane with an elliptic hole. The domain Σ is Σ^-, that is the region outside the unit circle γ ($|\zeta| \geq 1$). More precisely, the conformal transformation is defined on Σ^- by:

$$z = \omega(\zeta) = R(\zeta + \frac{m}{\zeta}) \qquad m < 1 \qquad [4.2]$$

The image of the unit circle is the ellipse, in which horizontal (a) and vertical (b) semi-axes are given by:

$$a = R(1+m); \quad b = R(1-m) \qquad [4.3]$$

To begin with (section 4.1.1), let us assume that the resultant force acting on the ellipse boundary is equal to 0 and that the stresses vanish at infinity. The general case is addressed in section 4.1.2.

4.1.1. $F_x = F_y = \Gamma = \Gamma' = 0$

The functional equation [1.66] in which the unknown is $\phi(\zeta)$ as well as the expression [1.69] of $\psi(\zeta)$ involve the following quantities defined on the unit circle γ:

$$|\sigma| = 1: \quad \frac{\omega(\sigma)}{\overline{\omega'(\sigma)}} = \frac{1}{\sigma}\frac{\sigma^2+m}{1-m\sigma^2}; \quad \frac{\overline{\omega(\sigma)}}{\omega'(\sigma)} = \sigma\,\frac{1+m\sigma^2}{\sigma^2-m} \qquad [4.4]$$

It is observed that $\omega(\sigma)/\overline{\omega'(\sigma)}$ is the boundary value on γ of the function:

$$\zeta \to \frac{1}{\zeta}\frac{\zeta^2+m}{1-m\zeta^2}$$

$\phi(\zeta)$ being holomorphic on Σ^- including the point $z = \infty$, it can be represented by a series expansion of form [1.95] (see section 1.7.3). Accordingly, we have:

$$\sigma \in \gamma : \quad \overline{\phi'(\sigma)} = \sigma \sum_{-\infty}^{-1} k\overline{a}_k \sigma^{-k}$$

This implies that $\overline{\phi'(\sigma)}$ is the boundary value on γ of the function:

$$\zeta \to \zeta \sum_{-\infty}^{-1} k\overline{a}_k \zeta^{-k}$$

Combining these remarks, the function

$$\frac{\omega(\sigma)}{\overline{\omega'(\sigma)}} \overline{\phi'(\sigma)}$$

of the second integral of [1.66] is the boundary value of function $F(\zeta)$ defined as:

$$F(\zeta) = \frac{\zeta^2 + m}{1 - m\zeta^2} \sum_{-\infty}^{-1} k\overline{a}_k \zeta^{-k} \quad [4.5]$$

Recalling that $m < 1$, it appears that $F(\zeta)$ is holomorphic on Σ^+ ($|\zeta| < 1$). It follows that this second integral of [1.66] is equal to 0, which yields:

$$|\zeta| > 1 : \quad \phi(\zeta) = -\frac{1}{2i\pi} \int_\gamma \frac{f \, d\sigma}{\sigma - \zeta} \quad [4.6]$$

In other words, we have an explicit expression of the complex potential $\phi(\zeta)$ as a function of the loading associated with f. Similarly, consider the function

$$\sigma \to \frac{\overline{\omega(\sigma)}}{\omega'(\sigma)}\phi'(\sigma)$$

which is involved in the second integral of [1.69]. It is the boundary value on γ of the function

$$\zeta \to \frac{1+m\zeta^2}{\zeta^2-m}\sum_{-\infty}^{-1} k a_k \zeta^k = \zeta\frac{1+m\zeta^2}{\zeta^2-m}\phi'(\zeta)$$

which is clearly holomorphic on Σ^-, including the point $z=\infty$, and vanishing at infinity. Owing to theorem [1.94], we obtain:

$$\frac{1}{2i\pi}\int_\gamma \frac{\overline{\omega(\sigma)}}{\omega'(\sigma)}\frac{\phi'(\sigma)}{\sigma-\zeta}\,d\sigma = -\zeta\frac{1+m\zeta^2}{\zeta^2-m}\phi'(\zeta)$$

This eventually yields the following expression of $\psi(\zeta)$ in Σ^-:

$$\psi(\zeta) = -\frac{1}{2i\pi}\int_\gamma \frac{\overline{f}}{\sigma-\zeta}\,d\sigma - \zeta\frac{1+m\zeta^2}{\zeta^2-m}\phi'(\zeta) \qquad [4.7]$$

4.1.2. $F_x, F_y, \Gamma, \Gamma' \neq 0$

Let us repeat the reasoning of section 1.7.3. The complex potentials $\phi(\zeta)$ and $\psi(\zeta)$ are given by [1.70]. The function f is replaced by the function f_o of [1.71] defined on γ. Owing to [4.4], f_o takes the form:

$$f_o = f - \frac{\overline{\Gamma'}R}{\sigma} - \Gamma R\left(\sigma + \frac{1}{\sigma}\frac{\sigma^2+m}{1-m\sigma^2}\right)$$
$$+\frac{F_x - iF_y}{2\pi(1+\kappa)}\frac{\sigma^2+m}{1-m\sigma^2} + \frac{F_x+iF_y}{2\pi}\log\sigma \qquad [4.8]$$

The complex functions $\phi_o(\zeta)$ and $\psi_o(\zeta)$ of [1.70] are derived from [4.6] and [4.7] obtained in the earlier section:

$$\phi_o(\zeta) = -\frac{1}{2i\pi}\int_\gamma \frac{f_o\,d\sigma}{\sigma-\zeta};$$
$$\psi_o(\zeta) = -\frac{1}{2i\pi}\int_\gamma \frac{\overline{f}_o}{\sigma-\zeta}\,d\sigma - \zeta\frac{1+m\zeta^2}{\zeta^2-m}\phi_o'(\zeta) \qquad [4.9]$$

4.1.3. *Elliptic cavity in a plane subjected to a remote stress state at infinity*

For illustrative purposes, let us define the loading by a given asymptotic stress state σ^∞. The edge of the elliptic cavity is free of stress. By application of [1.71], we first obtain:

$$f_o = -\frac{\overline{\Gamma'}R}{\sigma} - \Gamma R\left(\sigma + \frac{1}{\sigma}\frac{\sigma^2+m}{1-m\sigma^2}\right) \qquad [4.10]$$

The use of [4.9] for the derivation of $\phi_o(\zeta)$ involves the following integrals:

$$|\zeta|>1: \quad \frac{1}{2i\pi}\int_\gamma \frac{1}{\sigma}\frac{\sigma^2+m}{1-m\sigma^2}\frac{d\sigma}{\sigma-\zeta} = -\frac{m}{\zeta};$$
$$\frac{1}{2i\pi}\int_\gamma \frac{d\sigma}{\sigma(\sigma-\zeta)} = -\frac{1}{\zeta} \qquad [4.11]$$

It is then readily seen that:

$$\phi_o(\zeta) = -\frac{R}{\zeta}(\overline{\Gamma'}+m\Gamma); \quad \phi(\zeta) = \Gamma R\zeta - \frac{R}{\zeta}(\overline{\Gamma'}+m\Gamma) \qquad [4.12]$$

From [4.10], we obtain:

$$\overline{f}_o = -\Gamma' R\sigma - \Gamma R\left(\frac{1}{\sigma} + \sigma\frac{m\sigma^2+1}{\sigma^2-m}\right) \qquad [4.13]$$

The use of [4.9] for the derivation of $\psi_o(\zeta)$ in turn involves the following integral ($m < 1$):

$$\frac{1}{2i\pi}\int_\gamma \sigma\frac{1+m\sigma^2}{\sigma^2-m}\frac{d\sigma}{\sigma-\zeta} = (1+m^2)\frac{\zeta}{m-\zeta^2}$$

which eventually yields:

$$\psi_o(\zeta) = 2\Gamma R\frac{\zeta(1+m^2)}{m-\zeta^2} + \overline{\Gamma'}R\frac{1+m\zeta^2}{\zeta(m-\zeta^2)};$$

$$\psi_(\zeta) = \psi_o(\zeta) + \Gamma' R\zeta \qquad [4.14]$$

4.1.4. *Stress intensity factors*

The limit case $m \to 1$, which asymptotically corresponds to the Griffith crack, is now considered. [4.3] now becomes:

$$a = 2R; \quad b = 0 \qquad [4.15]$$

[4.12] and [4.14] take the following forms:

$$\phi_o(\zeta) = -\frac{R}{\zeta}(\overline{\Gamma'}+\Gamma); \quad \phi(\zeta) = \Gamma R\zeta - \frac{R}{\zeta}(\overline{\Gamma'}+\Gamma) \qquad [4.16]$$

and

$$\psi_o(\zeta) = 4\Gamma R\frac{\zeta}{1-\zeta^2} + \overline{\Gamma'}R\frac{1+\zeta^2}{\zeta(1-\zeta^2)}; \ \psi_(\zeta) = \psi_o(\zeta) + \Gamma' R\zeta \qquad [4.17]$$

In view of the determination of the stress intensity factors,[1] we make use of [1.28] and [1.29]. It is emphasized

1 A direct derivation of the stress intensity factors from the derivative $\phi'(z)$ is also possible [BUI 78, BUI 06, LIE 68].

that the latter involve the complex functions $\phi_1(z)$ and $\psi_1(z)$ (see equation [1.51]):

$$\sigma_{xx} = \phi_1'(z) + \overline{\phi_1'(z)} - \frac{1}{2}\left(\psi_1'(z) + \overline{\psi_1'(z)} + \overline{z}\phi''_1(z) + z\overline{\phi''_1(z)}\right) \qquad [4.18]$$

$$\sigma_{yy} = \phi_1'(z) + \overline{\phi_1'(z)} + \frac{1}{2}\left(\psi_1'(z) + \overline{\psi_1'(z)} + \overline{z}\phi''_1(z) + z\overline{\phi''_1(z)}\right) \qquad [4.19]$$

$$\sigma_{xy} = \frac{1}{2i}\left(\psi_1'(z) - \overline{\psi_1'(z)} + \overline{z}\phi''_1(z) - z\overline{\phi''_1(z)}\right) \qquad [4.20]$$

The expressions of the derivatives of $\phi_1(z)$ and $\psi_1(z)$ are given as functions of ζ:

$$\phi_1'(z) = \frac{\phi'(\zeta)}{\omega'(\zeta)} = \Gamma\left(\frac{\zeta^2+1}{\zeta^2-1}\right) + \overline{\Gamma'}\frac{1}{\zeta^2-1} \qquad [4.21]$$

$$\phi''_1(z) = -\frac{2\zeta^3}{(\zeta^2-1)^3}\frac{\overline{\Gamma'}+2\Gamma}{R} \qquad [4.22]$$

$$\psi_1'(z) = \frac{\psi'(\zeta)}{\omega'(\zeta)} = 4\Gamma\frac{\zeta^2(\zeta^2+1)}{(\zeta^2-1)^3} + \Gamma'\frac{\zeta^2}{\zeta^2-1} + \overline{\Gamma'}\frac{\zeta^4+4\zeta^2-1}{(\zeta^2-1)^3} \qquad [4.23]$$

On the axis, $y = 0$ ($\zeta \in \mathbb{R}$):

$$\sigma_{yy} = 2\Gamma\frac{\zeta^2+1}{\zeta^2-1} + \frac{\Gamma'+\overline{\Gamma'}}{\zeta^2-1} + \frac{\Gamma'+\overline{\Gamma'}}{2} \qquad [4.24]$$

Using [1.49], we obtain:

$$\sigma_{yy} = \sigma_{yy}^{\infty}\left(1 + \frac{2}{\zeta^2-1}\right) \qquad [4.25]$$

from which the stress intensity factor K_I is easily determined

$$K_I = \lim_{z \to a} \sqrt{2(z-a)}\sigma_{yy} = \sigma_{yy}^{\infty}\sqrt{a} \quad [4.26]$$

in which z was replaced by $\omega(\zeta) = a(\zeta + 1/\zeta)/2$ (see [4.2] and [4.15] with $m = 1$). Similarly, we get

$$\sigma_{xx} = 2\Gamma\frac{\zeta^2+1}{\zeta^2-1} + \frac{\Gamma' + \overline{\Gamma'}}{\zeta^2-1} - \frac{\Gamma' + \overline{\Gamma'}}{2} \quad [4.27]$$

which yields:

$$\sigma_{xx} = \sigma_{xx}^{\infty} + \sigma_{yy}^{\infty}\frac{2}{\zeta^2-1} \quad [4.28]$$

and furthermore:

$$\lim_{z \to a} \sqrt{2(z-a)}\sigma_{xx} = \sigma_{yy}^{\infty}\sqrt{a} = K_I \quad [4.29]$$

Finally, from [4.20], we get:

$$\sigma_{xy} = \frac{1}{2i}(\Gamma' - \overline{\Gamma'})\frac{\zeta^2+1}{\zeta^2-1} \quad [4.30]$$

or

$$\sigma_{xy} = \sigma_{xy}^{\infty}\frac{\zeta^2+1}{\zeta^2-1} \quad [4.31]$$

Accordingly, the stress intensity factor K_{II} reads:

$$K_{II} = \lim_{z \to a} \sqrt{2(z-a)}\sigma_{xy} = \sigma_{xy}^{\infty}\sqrt{a} \quad [4.32]$$

The present approach retrieves the results of [3.67] and [3.93].

Let us now consider the displacement discontinuity, that is the displacement of the upper crack lip with respect to that of the lower crack lip. Recalling [4.2], it is readily seen that the upper lip is the image by ω of the points $\sigma = e^{i\theta}$ with $\theta \in [0, \pi]$ (upper half-unit circle), whereas the lower lip is the image of the points $\sigma = e^{i\theta}$ with $\theta \in [\pi, 2\pi]$ (lower half-unit circle). Two points z^+ and z^- located at the same abscissa $x = 2R\cos\theta$, respectively, on the upper and lower lips are therefore the images of $\sigma^+ = e^{i\theta}$ and of $\sigma^- = e^{i(2\pi-\theta)}$. It follows that $\sigma^+ = \overline{\sigma^-}$. Recalling [1.39], and owing to [4.16]–[4.17], we have:

$$2\mu\left([\![\xi_x]\!] + i[\![\xi_y]\!]\right) = 2R\sin\theta(1+\kappa)\left(\sigma_{xy}^{\infty} + i\sigma_{yy}^{\infty}\right) \qquad [4.33]$$

Since $z^{\pm} = a\cos\theta$ $(a = 2R)$ and $\kappa = 3 - 4\nu$, we have furthermore:

$$[\![\xi_x]\!] + i[\![\xi_y]\!] = \frac{4(1-\nu^2)}{E}\left(\sigma_{xy}^{\infty} + i\sigma_{yy}^{\infty}\right)\sqrt{a^2 - x^2} \qquad [4.34]$$

We note that results [3.70] and [3.96] are retrieved.

4.1.5. *Some remarks on unilateral contact*

All the developments of the previous section implicitly assume that the crack is opened. This means that the free-of-stress condition on the crack lips is satisfied. More precisely, the concept of an opened crack can be defined by the following conditions:

$$\sigma_{xy} = \sigma_{yy} = 0 : \quad [\![\xi_y]\!] \geq 0 \qquad [4.35]$$

Owing to [4.34], this implies that $\sigma_{yy}^{\infty} \geq 0$, otherwise crack closure takes place. This suggests introducing the concept of a frictionless closed crack, which is a model of unilateral contact:

$$\sigma_{xy} = 0 \quad : \quad \sigma_{yy} \leq 0 : \quad [\![\xi_y]\!] = 0 \qquad [4.36]$$

Clearly enough, the transition between the two regimes is defined by the conditions:

$$\sigma_{xy} = \sigma_{yy} = 0 : \quad [\![\xi_y]\!] = 0 \qquad [4.37]$$

In the case of uniaxial compression at infinity in the direction normal to the crack ($\sigma_{yy}^{\infty} \leq 0$ and $\sigma_{xy}^{\infty} = 0$), the crack does not introduce any perturbation: a uniform stress field is preserved and the displacement discontinuity is 0. We will see in section 4.2.6.4 that this result is still valid in the anisotropic case. In turn, when the loading at infinity is a pure shear stress state ($\sigma_{xy}^{\infty} \neq 0$, $\sigma_{yy}^{\infty} = 0$), the sliding of the crack lips on one another is strictly identical to that obtained with the same loading in the case of an opened crack. Indeed, it is readily seen that the corresponding solution is compatible with conditions [4.35] as well as [4.36]. For closed cracks, we can summarize these conclusions by:

$$[\![\xi_x]\!] + i[\![\xi_y]\!] = \frac{4(1-\nu^2)}{E}\sigma_{xy}^{\infty}\sqrt{a^2 - x^2} \qquad [4.38]$$

which is identical to [4.34] in the case of pure shear at infinity. In contrast, in the case of remote shear stresses, different displacement discontinuities will be obtained for opened or closed cracks in the anisotropic solid.

4.2. Infinite plane with elliptic hole: the anisotropic case

4.2.1. *General properties*

In section 4.1, we consider an infinite plane with an elliptic cavity[2] with boundary L_z. Again, the idea is to use the

2 An alternative approach based on complex potentials within the Griffith crack model can be found in [AZH 00].

conformal transformation [4.2]. With the notations of [4.3], we get:

$$z = \frac{a+b}{2}\zeta + \frac{a-b}{2}\frac{1}{\zeta} \qquad [4.39]$$

The studied domain appears as an image of the region outside the unit circle (domain Σ^- defined by $|\zeta| \geq 1$). The point z is the image of point $\zeta = e^{i\theta}$ on the unit circle γ due to the conformal transformation [4.39]. In turn, $z_i = x + \mu_i y$ is the image of point z by the transformation:

$$z_i = a\cos\theta + b\mu_i \sin\theta \qquad [4.40]$$

which also reads:

$$i = 1, 2;\ \zeta \in \gamma : z_i = \frac{a - i\mu_i b}{2}\zeta + \frac{a + i\mu_i b}{2}\frac{1}{\zeta} \qquad [4.41]$$

More generally, two conformal transformations of region Σ^- outside the unit circle are considered:

$$i = 1, 2;\ \zeta_i \in \Sigma^- : \quad z_i = \omega_i(\zeta_i) = \frac{a - i\mu_i b}{2}\zeta_i + \frac{a + i\mu_i b}{2}\frac{1}{\zeta_i} \qquad [4.42]$$

This definition ensures the following property: according to [4.41], if a point $z = x + iy$ is located on the boundary L_z of the elliptic cavity, then z, $z_1 = x + \mu_1 y$ and $z_2 = x + \mu_2 y$ are related to the same ζ located on γ by [4.39] and [4.42], respectively:

$$z \in L_z : \quad \zeta = \zeta_1 = \zeta_2 \qquad [4.43]$$

Following the same reasoning as in section 1.7.1 (see equation [1.51]), a change of variable is introduced in the potentials $\phi_1(z_1)$ and $\phi_2(z_2)$

$$\varphi_1(\zeta_1) = \phi_1\left(\omega_1(\zeta_1)\right); \quad \varphi_2(\zeta_2) = \phi_2\left(\omega_2(\zeta_2)\right) \qquad [4.44]$$

Taking advantage of relations in [4.43], the boundary condition [1.87] on the boundary of the elliptic cavity reads:

$$\zeta \in \gamma \begin{cases} 2\mathcal{R}e\,(\mu_1\varphi_1(\zeta) + \mu_2\varphi_2(\zeta)) = F_x + \text{Const.} \\ 2\mathcal{R}e\,(\varphi_1(\zeta) + \varphi_2(\zeta)) = -F_y + \text{Const.} \end{cases} \qquad [4.45]$$

Now, we successively examine the response of the system to uniaxial traction and shear at infinity. In both cases, bearing [1.82] in mind, the existence of an asymptotic stress at infinity suggests that the potentials $\phi_1(z_1)$ and $\phi_2(z_2)$ should comprise:

– a linear part with respect to z_1 (respectively, z_2), which accounts for the asymptotic stress state;

– a holomorphic part at infinity.

These potentials are therefore sought in the form [LEK 63]:

$$\phi_i(z_i) = \Gamma_i z_i + \sum_{k \geq 0} a_k^{(i)} z_i^{-k} \qquad [4.46]$$

The Γ_i are related to the asymptotic stress state by:

$$\begin{aligned} \sigma_{xx}^{\infty} &= 2\mathcal{R}e\left(\mu_1^2\Gamma_1 + \mu_2^2\Gamma_2\right) \\ \sigma_{yy}^{\infty} &= 2\mathcal{R}e\,(\Gamma_1 + \Gamma_2) \\ \sigma_{xy}^{\infty} &= -2\mathcal{R}e\,(\mu_1\Gamma_1 + \mu_2\Gamma_2) \end{aligned} \qquad [4.47]$$

The change of variable [4.44] yields the following:

$$\varphi_i(\zeta_i) = \Gamma_i \omega_i(\zeta_i) + \tilde{\varphi}_i(\zeta_i) \quad \text{with} \quad \tilde{\varphi}_i(\zeta_i) = \sum_{k \geq 0} b_k^{(i)} \zeta_i^{-k} \qquad [4.48]$$

or

$$\varphi_i(\zeta_i) = \Gamma_i \left(\frac{a - i\mu_i b}{2} \zeta_i + \frac{a + i\mu_i b}{2} \frac{1}{\zeta_i} \right) + \tilde{\varphi}_i(\zeta_i) \qquad [4.49]$$

Using [4.49] in [1.87], the free-of-stress boundary condition on the cavity wall yields the following requirements[3]:

$$\zeta \in \gamma \begin{cases} 2\mathcal{R}e\left(\mu_1\tilde{\varphi}_1(\zeta) + \mu_2\tilde{\varphi}_2(\zeta)\right) = -2\mathcal{R}e\left(\mu_1\Gamma_1\omega_1(\zeta) + \mu_2\Gamma_2\omega_2(\zeta)\right) \\ 2\mathcal{R}e\left(\tilde{\varphi}_1(\zeta) + \tilde{\varphi}_2(\zeta)\right) = -2\mathcal{R}e\left(\Gamma_1\omega_1(\zeta) + \Gamma_2\omega_2(\zeta)\right) \end{cases} \quad [4.50]$$

Owing to [4.42], the quantities appearing in the right-hand side are as follows:

$$\begin{aligned} &-2\mathcal{R}e\left(\mu_1\Gamma_1\omega_1(\zeta) + \mu_2\Gamma_2\omega_2(\zeta)\right) = \\ &\qquad \mathcal{R}e\left(-a(\mu_1\Gamma_1 + \mu_2\Gamma_2)(\zeta + \frac{1}{\zeta}) + b(\mu_1^2\Gamma_1 + \mu_2^2\Gamma_2)(\zeta - \frac{1}{\zeta})i\right) \\ &-2\mathcal{R}e\left(\Gamma_1\omega_1(\zeta) + \Gamma_2\omega_2(\zeta)\right) = \\ &\qquad \mathcal{R}e\left(-a(\Gamma_1 + \Gamma_2)(\zeta + \frac{1}{\zeta}) + b(\mu_1\Gamma_1 + \mu_2\Gamma_2)(\zeta - \frac{1}{\zeta})i\right) \end{aligned} \quad [4.51]$$

4.2.2. *Complex potentials for an elliptic cavity in the presence of traction at infinity*

The asymptotic stress state is defined as follows:

$$\sigma^{\infty} = \sigma^{\infty}_{yy}\underline{e}_y \otimes \underline{e}_y \quad [4.52]$$

Owing to [1.82], we observe that the Γ_i of [4.46] are subjected to the conditions as follows:

$$\begin{aligned} &\mathcal{R}e(\mu_1^2\Gamma_1 + \mu_2^2\Gamma_2) = 0 \\ &\mathcal{R}e(\Gamma_1 + \Gamma_2) = \frac{\sigma^{\infty}_{yy}}{2} \\ &\mathcal{R}e(\mu_1\Gamma_1 + \mu_2\Gamma_2) = 0 \end{aligned} \quad [4.53]$$

It follows from [4.51] that

$$(\forall\zeta \in \gamma) \quad -2\mathcal{R}e\left(\mu_1\Gamma_1\omega_1(\zeta) + \mu_2\Gamma_2\omega_2(\zeta)\right) = 0 \quad [4.54]$$

3 The value given to the constants is 0.

as well as

$$(\forall\zeta\in\gamma)\quad -2\mathcal{R}e\left(\Gamma_1\omega_1(\zeta)+\Gamma_2\omega_2(\zeta)\right)=-a\frac{\sigma_{yy}^{\infty}}{2}\left(\zeta+\frac{1}{\zeta}\right)\qquad [4.55]$$

The problem consists of finding the $b_k^{(i)}$ $(i=1,2)$ meeting [4.50]:

$$\forall\zeta=e^{i\theta}\quad\begin{cases}2\mathcal{R}e\left(\sum_{k\geq 0}(\mu_1 b_k^{(1)}+\mu_2 b_k^{(2)})\zeta^{-k}\right)=0\\ 2\mathcal{R}e\left(\sum_{k\geq 0}(b_k^{(1)}+b_k^{(2)})\zeta^{-k}\right)=-a\frac{\sigma_{yy}^{\infty}}{2}\left(\zeta+\frac{1}{\zeta}\right)\end{cases}\qquad [4.56]$$

It is readily seen that $b_k^{(i)}=0$ if $k\neq 1$ and

$$b_1^{(1)}=a\sigma_{yy}^{\infty}\frac{\mu_2}{2(\mu_1-\mu_2)};\quad b_1^{(2)}=-a\sigma_{yy}^{\infty}\frac{\mu_1}{2(\mu_1-\mu_2)}\qquad [4.57]$$

This completes the determination of the potentials $\tilde{\varphi}_i(\zeta_i)$ of [4.48].

4.2.3. *Complex potentials for an elliptic cavity in the case of shear at infinity*

The asymptotic stress state is now defined as:

$$\sigma^{\infty}=\sigma_{xy}^{\infty}\left(\underline{e}_x\otimes\underline{e}_y+\underline{e}_y\otimes\underline{e}_x\right)\qquad [4.58]$$

From [1.82], it is found that the Γ_i introduced in [4.46] are subjected to the following conditions:

$$\begin{aligned}&\mathcal{R}e(\mu_1^2\Gamma_1+\mu_2^2\Gamma_2)=0\\&\mathcal{R}e(\Gamma_1+\Gamma_2)=0\\&\mathcal{R}e(\mu_1\Gamma_1+\mu_2\Gamma_2)=-\frac{\sigma_{xy}^{\infty}}{2}\end{aligned}\qquad [4.59]$$

From [4.51], it follows that

$$(\forall \zeta \in \gamma) - 2\mathcal{R}e\left(\mu_1 \Gamma_1 \omega_1(\zeta) + \mu_2 \Gamma_2 \omega_2(\zeta)\right) = a\frac{\sigma_{xy}^{\infty}}{2}\left(\zeta + \frac{1}{\zeta}\right) \quad [4.60]$$

as well as

$$(\forall \zeta \in \gamma) - 2\mathcal{R}e\left(\Gamma_1 \omega_1(\zeta) + \Gamma_2 \omega_2(\zeta)\right) = -b\frac{\sigma_{xy}^{\infty}}{2}i\left(\zeta - \frac{1}{\zeta}\right) \quad [4.61]$$

Again, the problem consists of finding the $b_k^{(i)}$ $(i = 1, 2)$ meeting [4.50]:

$$\forall \zeta = e^{i\theta} \begin{cases} 2\mathcal{R}e\left(\sum_{k \geq 0}(\mu_1 b_k^{(1)} + \mu_2 b_k^{(2)})\zeta^{-k}\right) = a\frac{\sigma_{xy}^{\infty}}{2}\left(\zeta + \frac{1}{\zeta}\right) \\ 2\mathcal{R}e\left(\sum_{k \geq 0}(b_k^{(1)} + b_k^{(2)})\zeta^{-k}\right) = -b\frac{\sigma_{xy}^{\infty}}{2}i\left(\zeta - \frac{1}{\zeta}\right) \end{cases} \quad [4.62]$$

It is found that $b_k^{(i)} = 0$ if $k \neq 1$ and

$$b_1^{(1)} = \frac{\sigma_{xy}^{\infty}}{2}\frac{a - i\mu_2 b}{\mu_1 - \mu_2}; \quad b_1^{(2)} = -\frac{\sigma_{xy}^{\infty}}{2}\frac{a - i\mu_1 b}{\mu_1 - \mu_2} \quad [4.63]$$

4.2.4. *Stresses in the limit case* $b \to 0$ *(crack)*

In view of the derivation of the stress from [1.82], we need to determine $\phi_j'(z_j)$ and $\varphi_j'(\zeta_j)$:

$$\phi_j'(z_j) = \frac{\varphi_j'(\zeta_j)}{\omega_j'(\zeta_j)} = \Gamma_j + \frac{\tilde{\varphi}_j'(\zeta_j)}{\omega_j'(\zeta_j)} \quad [4.64]$$

where [4.44] and [4.48] have been used. Let us now consider the limit case $b \to 0$. Recalling [4.42], it is found that:

$$z_j = \omega_j(\zeta_j) \approx \frac{a}{2}\left(\zeta_j + \frac{1}{\zeta_j}\right) \qquad [4.65]$$

It is convenient to introduce $F_j(\zeta_j)$ defined by:

$$\phi'_j(z_j) = \Gamma_j + F_j(\zeta_j) \quad \text{with} \quad F_j(\zeta_j) = \frac{2\tilde{\varphi}'_j(\zeta_j)}{a(1 - 1/\zeta_j^2)} \qquad [4.66]$$

Owing to [1.82] and [4.47], it is readily seen that:

$$\begin{aligned} \sigma_{xx} &= \sigma_{xx}^{\infty} + 2\mathcal{R}e\left(\mu_1^2 F_1(\zeta_1) + \mu_2^2 F_2(\zeta_2)\right) \\ \sigma_{yy} &= \sigma_{yy}^{\infty} + 2\mathcal{R}e\left(F_1(\zeta_1) + F_2(\zeta_2)\right) \\ \sigma_{xy} &= \sigma_{xy}^{\infty} - 2\mathcal{R}e\left(\mu_1 F_1(\zeta_1) + \mu_2 F_2(\zeta_2)\right) \end{aligned} \qquad [4.67]$$

In view of the derivation of the stress intensity factors, we restrict ourselves to the limit on the x-axis, from the right of the crack tip ($x \to a^+$). This means that $y = 0$, so that $z = z_1 = z_2$ and $\zeta = \zeta_1 = \zeta_2$:

$$\zeta = \frac{x}{a} + \sqrt{(\frac{x}{a})^2 - 1} \qquad [4.68]$$

4.2.4.1. *Preliminary remark concerning the definition of loading modes*

The usual definition of modes I and II refers to the displacement discontinuity across the crack lips. A difficulty arises in the anisotropic case, in so far as even simple remote loadings at infinity, such as normal traction or shear, induce mixed kinematic response involving both normal and tangential displacement discontinuities. For this reason, we hereafter refer to mode I (respectively, mode II) as a loading defined by remote normal (respectively, shear) stresses at infinity.

4.2.4.2. *Stress intensity factors in mode I*

Owing to the results in section 4.2.2, the potentials of [4.48] read:

$$\tilde{\varphi}_1(\zeta_1) = \frac{b_1^{(1)}}{\zeta_1}; \quad \tilde{\varphi}_2(\zeta_2) = \frac{b_1^{(2)}}{\zeta_2} \qquad [4.69]$$

Under this form, the functions $F_j(\zeta_j)$ of [4.66] read:

$$F_j(\zeta_j) = \frac{2}{a}\frac{b_1^{(j)}}{1-\zeta_j^2} \qquad [4.70]$$

Owing to [4.67], [4.57] and [4.68], the stresses on the $x-$axis on the right of the crack tip are as follows:

$$\begin{aligned} \sigma_{xx} &= -\frac{2\sigma_{yy}^{\infty}}{\zeta^2-1}\mathcal{R}e(\mu_1\mu_2) \\ \sigma_{yy} &= \sigma_{yy}^{\infty}\left(1+\frac{2}{\zeta^2-1}\right) \\ \sigma_{xy} &= 0 \end{aligned} \qquad [4.71]$$

The stress σ_{yy} on the $x-$axis on the right of the crack tip has the same expression as in the isotropic case (see equation [4.25]). The stress intensity factor is therefore identical (see equation [4.26]):

$$K_I = \lim_{z\to a^+} \sqrt{2(z-a)}\sigma_{yy} = \sigma_{yy}^{\infty}\sqrt{a} \qquad [4.72]$$

where we have used

$$\lim_{z\to a^+} \frac{\sqrt{2(z-a)}}{\zeta^2-1} = \frac{\sqrt{a}}{2} \qquad [4.73]$$

In contrast, note that

$$\lim_{z \to a^+} \frac{\sigma_{xx}}{\sigma_{yy}} = -\mathcal{R}e(\mu_1 \mu_2) \qquad [4.74]$$

while the limit of this ratio is equal to 1 in the isotropic case.

4.2.4.3. *Stress intensity factors in mode II*

The reasoning is similar as in the earlier section. The complex potentials take form [4.69] with the coefficients $b_1^{(i)}$ given in [4.63]. [4.67] now yields:

$$\begin{aligned}
\sigma_{xx} &= -\frac{2\sigma_{xy}^{\infty}}{(\zeta^2 - 1)} \mathcal{R}e(\mu_1 + \mu_2) \\
\sigma_{yy} &= 0 \\
\sigma_{xy} &= \sigma_{xy}^{\infty} \left(1 + \frac{2}{\zeta^2 - 1}\right)
\end{aligned} \qquad [4.75]$$

As in the isotropic case, the stress intensity factor is as follows:

$$K_{II} = \lim_{z \to a^+} \sqrt{2(z-a)}\sigma_{xy} = \sigma_{xy}^{\infty}\sqrt{a} \qquad [4.76]$$

Still, unlike the isotropic case, σ_{xx} is generally *not* equal to 0. Indeed, we have

$$\lim_{z \to a^+} \sqrt{2(z-a)}\sigma_{xx} = -\sigma_{xy}^{\infty}\sqrt{a}\mathcal{R}e(\mu_1 + \mu_2) \qquad [4.77]$$

Note that

$$\mathcal{R}e(\mu_1 + \mu_2) = \frac{A_{16}}{A_{11}}$$

where coefficients A_{ij} are defined in section 1.8.

4.2.5. *Displacement discontinuities*

Plane strains are considered in this section. In this case, the displacements are related to the potentials $\phi_1(z_1)$ and $\phi_2(z_2)$ by [1.85]. These potentials can be determined from relations [4.44] and [4.48]. Moreover, in the limit case of a crack, the transformations are defined [4.65]. Finally, when the boundary of the crack is considered, we have $\zeta = \zeta_1 = \zeta_2$ (see equation [4.43]).

The analysis of the displacement discontinuity across the crack was presented in the isotropic case in section 4.1.4. It is recalled that two points located on each side of the crack at the same abscissa are associated with $\zeta^+ = e^{i\theta}$ and $\zeta^- = e^{i(2\pi-\theta)} = 1/\zeta^+$. Accordingly, we have:

$$\begin{aligned} [\![\xi_x]\!] &= 2\mathcal{R}e(p_1[\![\phi_1]\!] + p_2[\![\phi_2]\!]) \\ [\![\xi_y]\!] &= 2\mathcal{R}e(q_1[\![\phi_1]\!] + q_2[\![\phi_2]\!]) \end{aligned} \qquad [4.78]$$

In the case of mode I or mode II loadings, it is readily seen that

$$[\![\phi_j]\!](z) = [\![\tilde{\varphi}_j(\zeta)]\!] = -b_1^{(j)}(\zeta^+ - \frac{1}{\zeta^+}) = -2ib_1^{(j)} \sin\theta \qquad [4.79]$$

where the $b_1^{(i)}$ are given either by [4.57] or by [4.63]. It follows that:

$$\begin{aligned} [\![\xi_x]\!] &= -4\sin\theta\mathcal{R}e\big(i(p_1 b_1^{(1)} + p_2 b_1^{(2)})\big) \\ [\![\xi_y]\!] &= -4\sin\theta\mathcal{R}e\big(i(q_1 b_1^{(1)} + q_2 b_1^{(2)})\big) \end{aligned} \qquad [4.80]$$

where p_i and q_i were given in [1.84]. For the following, let us recall once more that, on the crack, we have $\zeta = e^{i\theta}$ and $z = x = a\cos\theta$ (see equations [4.43] and [4.65]). This also implies that $a\sin\theta = \sqrt{a^2 - x^2}$.

4.2.5.1. *Displacement discontinuity in mode I*

Let us consider mode I. Introducing [4.57] into [4.80], we obtain:

$$\begin{aligned} [\![\xi_x]\!] &= -2\sigma_{yy}^{\infty} b_{11}\sqrt{a^2 - x^2}\mathcal{R}e\big(i\mu_1\mu_2\big) \\ [\![\xi_y]\!] &= 2\sigma_{yy}^{\infty} b_{22}\sqrt{a^2 - x^2}\mathcal{R}e\big(i\frac{\mu_1 + \mu_2}{\mu_1\mu_2}\big) \end{aligned} \qquad [4.81]$$

The property $b_{22} > 0$ is a direct consequence of the definite positiveness of the compliance tensor $\mathbb{S}$. Moreover, recalling that the imaginary parts of μ_1 and μ_2 are positive, it is readily seen that:

$$\mathcal{R}e\big(i\frac{\mu_1 + \mu_2}{\mu_1\mu_2}\big) \geq 0 \qquad [4.82]$$

It follows that $[\![\xi_y]\!]$ has the same sign as σ_{yy}^{∞}, as expected physically. This sign must be positive for the opened crack model to be physically relevant (see section 4.1.5).

4.2.5.2. *Discontinuity of displacement in mode II*

Introducing [4.63] into [4.80], we obtain:

$$\begin{aligned} [\![\xi_x]\!] &= -2\sigma_{xy}^{\infty} b_{11}\sqrt{a^2 - x^2}\mathcal{R}e\big(i(\mu_1 + \mu_2)\big) \\ [\![\xi_y]\!] &= 2\sigma_{xy}^{\infty} b_{22}\sqrt{a^2 - x^2}\mathcal{R}e\big(\frac{i}{\mu_1\mu_2}\big) \end{aligned} \qquad [4.83]$$

As opposed to mode I, the sign of $[\![\xi_y]\!]$ cannot be predicted *a priori*, since the sign of

$$\mathcal{R}e\big(\frac{i}{\mu_1\mu_2}\big)$$

depends on the considered anisotropy.

In order to retrieve the results of the isotropic case, it suffices to take $\mu_1 = \mu_2 = i$.

4.2.6. *Closed cracks*

In this section, we again consider the model of frictionless closed crack introduced in section 4.1.5. It is recalled that the normal discontinuity displacement is prevented, and that the stress acting on the crack lips is purely normal. Physically speaking, this model is relevant only in the case of a compressive normal stress (see equation [4.35]). It is emphasized that we refer here to the *local* normal stress.

When such a crack is considered in an isotropic medium, it was shown previously that it does not induce a perturbation of the stress state prescribed at infinity when the latter is of the form $\sigma_{yy}^{\infty}\underline{e}_y \otimes \underline{e}_y$ with $\sigma_{yy}^{\infty} \leq 0$. In this case, the stress field is therefore uniform. It is also recalled that the solution in mode II is strictly identical irrespective of the fact that the crack is closed or not. Indeed, no normal discontinuity is observed in the case of an opened crack.

In turn, in an anisotropic medium, a closed crack does not induce any perturbation when the stress state at infinity is of the form $\sigma_{yy}^{\infty}\underline{e}_y \otimes \underline{e}_y$ with $\sigma_{yy}^{\infty} \leq 0$. However, in contrast to isotropy, a shear stress at infinity induces normal stresses acting on the crack lips. The solution corresponding to mode II is presented in the following section.

4.2.6.1. *Derivation of the complex potentials in mode II*

As compared to the case of opened cracks, the new feature is that the boundary conditions acting on the crack lips are mixed. The vector $\underline{e}_y$ being normal to the crack line, the conditions are that the local tangential stress σ_{xy} and the normal displacement discontinuity $[\![\xi_y]\!]$ are equal to 0. With the same reasoning as in section 4.2.3, it appears that the condition $\sigma_{xy} = 0$ is expressed by the first equation of [4.62]:

$$\forall\zeta = e^{i\theta} 2\mathcal{R}e\left(\sum_{k\geq 0}(\mu_1 b_k^{(1)} + \mu_2 b_k^{(2)})\zeta^{-k}\right) = a\frac{\sigma_{xy}^{\infty}}{2}\left(\zeta + \frac{1}{\zeta}\right) \quad [4.84]$$

which implies that

$$\begin{array}{ll} (\forall k > 1) & \mu_1 b_k^{(1)} + \mu_2 b_k^{(2)} = 0 \\ k = 1 & \mu_1 b_1^{(1)} + \mu_2 b_1^{(2)} = a\frac{\sigma_{xy}^{\infty}}{2} \end{array} \qquad [4.85]$$

In turn, the kinematic condition $[\![\xi_y]\!] = 0$ is expressed from [1.85]:

$$[\![\xi_y]\!] = 2\mathcal{R}e(q_1[\![\phi_1]\!] + q_2[\![\phi_2]\!])$$

and yields (see equation [4.79])

$$0 = \sum_{k\geq 1} \mathcal{R}e\big(i(q_1 b_k^{(1)} + q_2 b_k^{(2)})\big) \sin k\theta \qquad [4.86]$$

from which it is concluded that:

$$(\forall k \geq 1) \quad \mathcal{R}e\big(i(q_1 b_k^{(1)} + q_2 b_k^{(2)})\big) = 0 \qquad [4.87]$$

Conditions [4.85] and [4.87] are not sufficient to derive the coefficients $b_k^{(j)}$. We still have to express the continuity condition of the normal stress σ_{yy} across the crack. To do so, [4.66] and [4.67] are used. On the upper lip, it is recalled that $\zeta_1 = \zeta_2 = \zeta^+ = e^{i\theta}$, while $\zeta_1 = \zeta_2 = \zeta^- = 1/\zeta^+$ holds on the lower lip. This yields:

$$\sum_{k\geq 1} \mathcal{R}e\Big(\frac{\tilde{\varphi}_1'(\zeta^+)}{1 - 1/\zeta^{+2}} + \frac{\tilde{\varphi}_2'(\zeta^+)}{1 - 1/\zeta^{+2}}\Big)$$

$$= \sum_{k\geq 1} \mathcal{R}e\Big(\frac{\tilde{\varphi}_1'(\zeta^-)}{1 - 1/\zeta^{-2}} + \frac{\tilde{\varphi}_2'(\zeta^-)}{1 - 1/\zeta^{-2}}\Big) \qquad [4.88]$$

from which we have:

$$(\forall k \geq 1) \quad \mathcal{I}m\left(b_k^{(1)} + b_k^{(2)}\right) = 0 \qquad [4.89]$$

From the set of equations [4.85], [4.87] and [4.89], it is possible to derive the two sequences $b_k^{(j)}$ $(j = 1, 2)$. Indeed, it is found that $b_k^{(j)} = 0$ for $k > 1$, while:

$$b_1^{(1)} = \frac{a}{2}\sigma_{xy}^{\infty}\beta_2 \frac{(\alpha_1 - \alpha_2 - i(\beta_1 + \beta_2))(\alpha_1 + i\beta_1)}{(\beta_1\alpha_2^2 + \alpha_1^2\beta_2 + \beta_2\beta_1^2 + \beta_1\beta_2^2)(\alpha_1 - \alpha_2 + i\beta_1 - i\beta_2)} \qquad [4.90]$$

$$b_1^{(2)} = \frac{a}{2}\sigma_{xy}^{\infty}\beta_1 \frac{(\alpha_1 - \alpha_2 + i(\beta_1 + \beta_2))(\alpha_2 + i\beta_2)}{(\beta_1\alpha_2^2 + \alpha_1^2\beta_2 + \beta_2\beta_1^2 + \beta_1\beta_2^2)(\alpha_1 - \alpha_2 + i\beta_1 - i\beta_2)} \qquad [4.91]$$

with the notations:

$$\mu_1 = \alpha_1 + i\beta_1; \quad \mu_2 = \alpha_2 + i\beta_2 \quad (\alpha_j, \in \mathbb{R},\ \beta_j, \in \mathbb{R}^+) \qquad [4.92]$$

4.2.6.2. *Displacement discontinuity in mode II*

The methodology is identical to that of section 4.2.5 and yields [4.80] in which [4.90] and [4.91] are to be used, from which we obtain:

$$[\![\xi_x]\!] = -2\sigma_{xy}^{\infty} b_{11}\sqrt{a^2 - x^2}\,\beta_1\beta_2 \frac{\alpha_2^2 - 2\alpha_2\alpha_1 + \alpha_1^2 + \beta_1^2 + 2\beta_2\beta_1 + \beta_2^2}{\beta_1\alpha_2^2 + \alpha_1^2\beta_2 + \beta_2\beta_1^2 + \beta_1\beta_2^2} \qquad [4.93]$$

In view of comparing with the discontinuity obtained in the case of an opened crack, it is useful to write [4.93] in the following form:

$$[\![\xi_x]\!] = -2\sigma_{xy}^{\infty} b_{11}\sqrt{a^2 - x^2}\left(\mathcal{R}e(i(\mu_1+\mu_2)) + \mathcal{I}m(\mu_1\mu_2)\frac{\mathcal{R}e(\frac{i}{\mu_1\mu_2})}{\mathcal{R}e(i\frac{\mu_1+\mu_2}{\mu_1\mu_2})}\right) \qquad [4.94]$$

Let the subscripts *cl* and o refer to the closed crack and the opened crack, respectively. Owing to [4.83] and [4.94], it is readily seen that:

$$[\![\xi_x]\!]_{cl} - [\![\xi_x]\!]_o = -2\sigma_{xy}^{\infty} b_{11}\sqrt{a^2 - x^2}\left(\mathcal{I}m(\mu_1\mu_2)\frac{\mathcal{R}e(\frac{i}{\mu_1\mu_2})}{\mathcal{R}e(i\frac{\mu_1+\mu_2}{\mu_1\mu_2})}\right) \qquad [4.95]$$

Furthermore, we have

$$\mathcal{I}m(\mu_1\mu_2)\frac{\mathcal{R}e(\frac{i}{\mu_1\mu_2})}{\mathcal{R}e(i\frac{\mu_1+\mu_2}{\mu_1\mu_2})} = \frac{(\alpha_1\beta_2 + \alpha_2\beta_1)^2}{\beta_1(\alpha_2^2 + \beta_2^2) + \beta_1(\alpha_1^2 + \beta_1^2)} \qquad [4.96]$$

Recalling that $\beta_i \geq 0$, it appears that the tangential displacement is greater in the case of the opened crack than in that of the closed crack:

$$\left([\![\xi_x]\!]_{cl} - [\![\xi_x]\!]_o\right)\sigma_{xy}^{\infty} \leq 0 \qquad [4.97]$$

In other words, the crack medium loaded in mode II is stiffer in the case of a closed crack. We will look at this conclusion again in Chapter 9.

4.2.6.3. *Local normal stress induced in mode II*

The normal stress acting on the crack lips is determined from relation [4.67] in which $\zeta = e^{i\theta}$ is used:

$$\sigma_{yy} = \sigma_{yy}^{\infty} - \frac{4}{a}\mathcal{R}e\left(\frac{b_1^{(1)} + b_1^{(2)}}{\zeta^2 - 1}\right) = \sigma_{yy}^{\infty} + 2\frac{b_1^{(1)} + b_1^{(2)}}{a} \qquad [4.98]$$

This normal stress is uniform. It is convenient to put this result in the form:

$$\sigma_{yy} = \sigma_{yy}^{\infty} + \sigma_{xy}^{\infty}\frac{\mathcal{R}e(\frac{i}{\mu_1\mu_2})}{\mathcal{R}e(i\frac{\mu_1+\mu_2}{\mu_1\mu_2})} = \frac{\mathcal{L}(\sigma_{yy}^{\infty}, \sigma_{xy}^{\infty})}{\mathcal{R}e(i\frac{\mu_1+\mu_2}{\mu_1\mu_2})} \qquad [4.99]$$

where

$$\mathcal{L}(\sigma_{yy}^{\infty}, \sigma_{xy}^{\infty}) = \sigma_{yy}^{\infty}\mathcal{R}e(i\frac{\mu_1 + \mu_2}{\mu_1\mu_2}) + \sigma_{xy}^{\infty}\mathcal{R}e(\frac{i}{\mu_1\mu_2}) \qquad [4.100]$$

4.2.6.4. *The opened–closed crack transition*

As mentioned previously, consider a straight crack with unit normal $\underline{e}_y$. The condition of opened crack is expressed on the two components of the stress vector acting on the crack lips, associated with a (unilateral) kinematic condition:

$$\sigma_{yy} = \sigma_{xy} = 0; \quad [\![\xi_y]\!] \geq 0 \qquad [4.101]$$

On the other hand, the condition of frictionless closed crack is expressed on the tangential component of the stress vector acting on the crack lips, on the normal component of the displacement discontinuity, associated with a condition of compressive stress as follows:

$$\sigma_{xy} = 0; \quad [\![\xi_y]\!] = 0; \quad \sigma_{yy} \leq 0 \qquad [4.102]$$

The transition state is therefore defined by the three scalar conditions:

$$\sigma_{yy} = \sigma_{xy} = 0; \quad [\![\xi_y]\!] = 0 \qquad [4.103]$$

The purpose of this section is to confirm that the transition from closed to opened regimes (and vice versa) can be characterized in a unique way, either from the model of an opened crack, or from the model of a closed crack.

In the framework of the opened crack model, [4.81]–[4.83] yield:

$$[\![\xi_y]\!] = 2b_{22}\sqrt{a^2 - x^2}\mathcal{L}(\sigma_{yy}^{\infty}, \sigma_{xy}^{\infty}) \qquad [4.104]$$

The condition of transition, that is $[\![\xi_y]\!] = 0$, thus takes the form:

$$\mathcal{L}(\sigma_{yy}^{\infty}, \sigma_{xy}^{\infty}) = 0 \qquad [4.105]$$

Alternatively, if the model of closed crack is considered, the transition is defined by the condition $\sigma_{yy} = 0$. Recalling [4.99], the transition rule [4.105] is clearly retrieved. It can be concluded that the two approaches provide the same mathematical characterization of the transition.

Moreover, it is instructive to observe that the displacement discontinuity $[\![\xi_y]\!]$ predicted by the opened crack model and given by [4.104], on the one hand, and the normal stress σ_{yy} predicted by the closed crack model and given by [4.99], on the other hand, have identical signs (recall [4.82]). This emphasizes the fact that the two models can be simultaneously consistent *only* if $\mathcal{L}(\sigma_{yy}^{\infty}, \sigma_{xy}^{\infty}) = 0$, which is the transition rule [4.105]. Indeed, if $\mathcal{L}(\sigma_{yy}^{\infty}, \sigma_{xy}^{\infty}) > 0$, then the normal stress predicted by [4.99] would be positive (local traction on the crack lips), which is not physically relevant.

Similarly, if $\mathcal{L}(\sigma_{yy}^{\infty}, \sigma_{xy}^{\infty}) < 0$, the opened crack model would predict interpenetration ($[\![\xi_y]\!] < 0$), which again is not relevant.

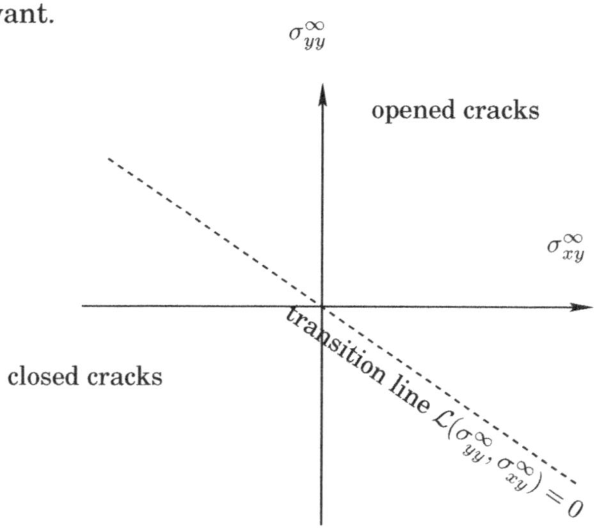

Figure 4.1. *Opened and closed cracks in the stress plane* $(\sigma_{xy}^{\infty}, \sigma_{yy}^{\infty})$

Figure 4.1 presents a partition of the plane $(\sigma_{xy}^{\infty}, \sigma_{yy}^{\infty})$ into two regions, which correspond to the domains of validity of opened versus closed crack models. Closed (respectively, opened[4]) cracks are characterized by $\mathcal{L}(\sigma_{yy}^{\infty}, \sigma_{xy}^{\infty}) < 0$ (respectively, $\mathcal{L}(\sigma_{yy}^{\infty}, \sigma_{xy}^{\infty}) > 0$). The transition between the regimes of closed and opened cracks is the straightline [4.105]. The sign of its slope depends on the material parameters that characterize the considered anisotropy. More precisely, the slope has the sign of

$$-\mathcal{R}e\left(\frac{i}{\mu_1\mu_2}\right)$$

which is the sign of $-(\alpha_1\beta_2 + \alpha_2\beta_1)$.

4 Recalling [4.82], it is readily seen that the half-axis $\sigma_{yy}^{\infty} > 0$ is included in the domain $\mathcal{L}(\sigma_{yy}^{\infty}, \sigma_{xy}^{\infty}) > 0$ corresponding to the regime of opened cracks.

4.3. Eshelby approach

In the framework of the elliptic crack model, the crack can be interpreted as an inhomogeneity of vanishing stiffness. With this point of view, the aim of the present section is to revisit the problem addressed in section 4.1 in the light of the solution of Eshelby's inhomogeneity problem. An isotropic elasticity of the solid is assumed in this section.

Eshelby's inhomogeneity problem has been introduced in section 2.4.4 in the framework of boundary conditions on the displacement field $\underline{\xi}$ at infinity, namely $\underline{\xi} \to \mathbf{E} \cdot \underline{z}$ for $|\underline{z}| \to \infty$. In other words, $\mathbf{E}$ can be interpreted as the strain tensor at infinity: $\mathbf{E} = \boldsymbol{\varepsilon}^{\infty}$. In contrast, section 4.1 (as well as Chapter 3) considers boundary conditions on the stress field $\boldsymbol{\sigma}$ at infinity, of the form $\sigma_{ij} \to \sigma_{ij}^{\infty}$ for $|\underline{z}| \to \infty$ $(i, j = 1, 2)$,[5] meaning that the asymptotic stress state $\boldsymbol{\sigma}^{\infty}$ is prescribed. In view of implementing the solution of Eshelby's inhomogeneity problem, we just have to determine the strain tensor $\boldsymbol{\varepsilon}^{\infty}$ associated with the considered $\boldsymbol{\sigma}^{\infty}$. The latter, of course, depends on the loading conditions (mode I or II).

4.3.1. *Mode I*

Mode I is defined by the conditions $\sigma_{11}^{\infty} = \sigma_{12}^{\infty} = 0$ and $\sigma_{22}^{\infty} = q$ $(q > 0)$. In addition, plane strain conditions imply that $\sigma_{13}^{\infty} = \sigma_{23}^{\infty} = 0$ and $\sigma_{33}^{\infty} = \nu(\sigma_{11}^{\infty} + \sigma_{22}^{\infty})$:

$$\boldsymbol{\sigma}^{\infty} = q(\underline{e}_2 \otimes \underline{e}_2 + \nu \underline{e}_3 \otimes \underline{e}_3) \qquad [4.106]$$

5 Note that z_1 and z_2 again denote the coordinates in the orthonormal Cartesian coordinate system.

The corresponding strain state at infinity is readily derived from the linear elastic state equation:

$$\varepsilon^{\infty} = \mathbb{C}^{-1} : \boldsymbol{\sigma}^{\infty} = (1-\nu^2)\frac{q}{E}\left(\underline{e}_2 \otimes \underline{e}_2 - \frac{\nu}{1-\nu}\underline{e}_1 \otimes \underline{e}_1\right) \quad [4.107]$$

We then conclude from section 2.4.4 that the elliptic crack (inhomogeneity I) undergoes a homogeneous strain ε^I given by [2.70] in which **E** is replaced by the above value of ε^{∞} and $\mathbb{C}^I = 0$. The components of $\mathbb{P}^I$ have been given in [2.105]. This yields:

$$\varepsilon^I = (1-\nu^2)\frac{q}{E}\left((\frac{2}{X}+1)\underline{e}_2 \otimes \underline{e}_2 - \underline{e}_1 \otimes \underline{e}_1\right) \quad [4.108]$$

Furthermore, the displacement field on the boundary of the ellipse is given by[6] $\underline{\xi}(\underline{z}) = \varepsilon^I \cdot \underline{z}$, with $z_2 = \epsilon X\sqrt{a^2 - z_1^2}$ ($\epsilon = \pm 1$):

$$\begin{aligned} \xi_1 &= -(1-\nu^2)\frac{q}{E}z_1 \\ \xi_2 &= \frac{q(1-\nu^2)\epsilon\sqrt{a^2-z_1^2}}{E}(2+X) \end{aligned} \quad [4.109]$$

Since the elliptic model assumes infinitesimal values of the aspect ratio X, the displacement jump across the upper and lower crack surfaces can be defined as:

$$[\![\underline{\xi}]\!] = = \lim_{X\to 0}\left(\underline{\xi}(z_1, +X\sqrt{a^2-z_1^2}) - \underline{\xi}(z_1, -X\sqrt{a^2-z_1^2})\right) \quad [4.110]$$

that is:

$$[\![\underline{\xi}]\!] = = 2\sqrt{a^2-z_1^2}\lim_{X\to 0}\left(X\varepsilon^I\right)\cdot \underline{e}_2 \quad [4.111]$$

---

6 In this case, note that symmetry arguments show that the deformation gradient in the elliptic inclusion has no skew-symmetric part.

Owing to [4.108], this yields:

$$[\![\underline{\xi}]\!] = 4(1-\nu^2)\frac{q}{E}\sqrt{a^2 - z_1^2}\underline{e}_2 \qquad [4.112]$$

This derivation of the displacement jump confirms the result of the Griffith crack model (see equation [3.70]).

In turn, we can easily derive the stress field on the boundary ∂I of the ellipse. To this end, denoting the unit vector normal (respectively, tangent) to ∂I by $\underline{n}$ (respectively, $\underline{t}$), we observe that the component ε_{tt} of the strain tensor and the components σ_{nt} and σ_{nn} of the stress tensor are continuous across ∂I:

$$\sigma_{nt} = \sigma_{nn} = 0; \quad \varepsilon_{tt} = \underline{t} \cdot \boldsymbol{\varepsilon}^I \cdot \underline{t} \qquad [4.113]$$

As far as the stresses are concerned, the only unknown is the component σ_{tt}. The latter is readily determined by the state equation in plane strain conditions using $\sigma_{nn} = 0$:

$$\sigma_{tt} = \frac{E}{1-\nu^2}\varepsilon_{tt} \qquad [4.114]$$

This result can applied to the crack tip $z_1 = a$, $z_2 = 0$, for which $\underline{t} = \underline{e}_2$:

$$\sigma_{22} = q(\frac{2}{X} + 1); \quad \sigma_{12} = \sigma_{11} = 0 \qquad [4.115]$$

It is interesting to compare this result to the stress state derived in the plane $z_2 = 0$ within the framework of the Griffith crack model (see equation [3.66]). As the ellipse tends toward the segment $|z_1| \leq a$, that is for $X \to 0$, we note from [4.115] that $\sigma_{22} \to \infty$. This is consistent with [3.66]. In contrast, in the elliptic model, the value of the normal stress σ_{11} is prescribed by the stress-free boundary condition, namely $\sigma_{11} = 0$, whereas it tends to infinity in the Griffith crack. However, using the results in section 4.1.3, it can be readily shown that the function $\sigma_{11}(z_1, 0)$ defined on $[a, \infty]$ reaches a maximum in the vicinity of $z_1 = a$ (crack tip). The value of this maximum tends to infinity, whereas the location of the peak tends to a^+ as the aspect ratio X tends to 0 ($m \to 1^-$). In this sense, the elliptic crack model is fully consistent with the Griffith crack model.

4.3.2. *Mode II*

Mode II is defined by the conditions $\sigma_{11}^{\infty} = \sigma_{22}^{\infty} = 0$ and $\sigma_{12}^{\infty} = q$. In addition, plane strain conditions imply that $\sigma_{i3}^{\infty} = 0$ ($i = 1, 2, 3$):

$$\boldsymbol{\sigma}^{\infty} = q(\underline{e}_2 \otimes \underline{e}_1 + \underline{e}_1 \otimes \underline{e}_2) \qquad [4.116]$$

The corresponding strain state at infinity reads:

$$\boldsymbol{\varepsilon}^{\infty} = \mathbb{C}^{-1} : \boldsymbol{\sigma}^{\infty} = (1+\nu)\frac{q}{E}\,(\underline{e}_2 \otimes \underline{e}_1 + \underline{e}_1 \otimes \underline{e}_2) \qquad [4.117]$$

The homogeneous strain $\boldsymbol{\varepsilon}^I$ in the elliptic crack is again given by [2.70] in which $\mathbf{E}$ is replaced by the above value of $\boldsymbol{\varepsilon}^{\infty}$ and $\mathbb{C}^I = 0$:

$$\boldsymbol{\varepsilon}^I = \frac{q}{E}\Big(\frac{1-\nu^2}{X}\,(\underline{e}_2 \otimes \underline{e}_1 + \underline{e}_1 \otimes \underline{e}_2) + \mathcal{O}(X^0)\Big) \qquad [4.118]$$

However, in the present case, the knowledge of $\boldsymbol{\varepsilon}^I$ is not sufficient for deriving the displacement field on the ellipse boundary [MUR 87, LAW 85]. Indeed, the latter takes the form $(\boldsymbol{\varepsilon}^I + \boldsymbol{w}^I) \cdot \underline{z}$, where $\boldsymbol{w}^I$ is the skewsymmetric part of the deformation gradient in the elliptic hole. It is therefore necessary to go back to the deformation gradient that has been derived in section 2.4.4. In [2.71], we note that ($\mathbb{C}^I = 0$):

$$\tau^I = (\mathbb{C}^I - \mathbb{C}_o) : \varepsilon^I = -q\frac{1-\nu}{X}(\underline{e}_2 \otimes \underline{e}_1 + \underline{e}_1 \otimes \underline{e}_2) + \mathcal{O}(X^0)) \quad [4.119]$$

which in turn yields:

$$\text{grad}\,\underline{\xi}^I = q\frac{1-\nu}{X}\mathbb{R}^I : (\underline{e}_2 \otimes \underline{e}_1 + \underline{e}_1 \otimes \underline{e}_2) + \mathcal{O}(1) \quad [4.120]$$

Restricting ourselves to the leading term, we obtain:

$$\xi_{1,2} = 2q\frac{1-\nu}{X}R_{1212} \quad [4.121]$$

Using the value of R_{1212} given in [2.106] yields the following:

$$\xi_{1,2} = 2\frac{q}{E}\frac{1-\nu^2}{X} \quad [4.122]$$

The (leading term of the) gradient of the displacement being uniform, the displacement field within the elliptic inclusion and on its boundary eventually reads[7]:

$$\underline{\xi} = \xi_{1,2} z_2 \underline{e}_1 + \xi_{2,1} z_1 \underline{e}_2 \quad [4.123]$$

Furthermore, in view of comparison with the solution of the 2D Griffith crack in mode II, the displacement jump is then derived according to definition [4.110]:

$$[\![\underline{\xi}]\!] = 2X\sqrt{a^2 - z_1^2}\,\xi_{1,2}\underline{e}_1 \quad [4.124]$$

7 Note that $\xi_{1,1} = \xi_{2,2} = 0$, since $\varepsilon^I_{11} = \varepsilon^I_{22} = 0$.

that is:

$$[\![\underline{\xi}]\!] = 4(1-\nu^2)\frac{\sigma_{12}^{\infty}}{E}\sqrt{a^2 - z_1^2}\underline{e}_1 \quad [4.125]$$

which confirms the result of the Griffith crack model (see equation [3.96]). Again, we can easily derive the unknown stress component σ_{tt} on the boundary ∂I of the ellipse using the fact that the component ε_{tt} of the strain tensor and the components σ_{nt} and σ_{nn} of the stress tensor are continuous across ∂I; [4.113] and [4.114] are still valid. At the crack tip $z_1 = a$, $z_2 = 0$, for which $\underline{t} = \underline{e}_2$, we obtain from [4.118] that $\varepsilon_{tt} = 0$, which in turn yields:

$$\sigma_{11} = \sigma_{12} = \sigma_{22} = 0 \quad [4.126]$$

This result is valid for any value of X, and in particular as the ellipse tends toward the segment $|z_1| \leq a$, that is for $X \to 0$. With regard to the shear component σ_{12}, which is prescribed by the stress-free boundary condition, there is of course an apparent contradiction with [3.92], because it tends to infinity within the framework of the Griffith crack model. This again emphasizes the importance of the geometrical model used for the crack, namely a segment in the Griffith model and an ellipse in the present model. A comment concerning σ_{12} similar to the one (concerning σ_{11}) at the end of section 4.3.1 can be formulated.

5

Griffith Crack in 3D

In sections 5.1 and 5.2 of this chapter, we try to find the displacement and the stress field solutions to the 3D Griffith circular crack problem. The geometrical model and the associated mathematics are the 3D counterpart of those presented in Chapter 3 in plane strain conditions. The crack appears as a plane surface[1] with unit normal vector $\underline{N}$, embedded in an infinite 3D homogeneous continuum. The two faces of the crack are free of stress and an asymptotic stress state $\boldsymbol{\sigma}^{\infty}$ is prescribed at infinity. In this chapter, the unit normal to the crack plane is $\underline{N} = \underline{e}_3$ in the cartesian coordinate system and $\underline{N} = \underline{e}_z$ in the cylindrical coordinate system.

As already stated in Chapter 3, this loading induces strain and stress singularities as well as displacement jump only if $\boldsymbol{\sigma}^{\infty} \cdot \underline{N} \neq 0$. Otherwise, the stress field is uniform and equal to $\boldsymbol{\sigma}^{\infty}$. We therefore first consider a mode I loading of the form $\boldsymbol{\sigma}^{\infty} = q\underline{N} \otimes \underline{N}$ (section 5.1). Next, considering a unit normal vector $\underline{T}$ in the crack plane ($\underline{T} \cdot \underline{N} = 0$), we derive the response to a shear loading defined by $\boldsymbol{\sigma}^{\infty} = q(\underline{N} \otimes \underline{T} + \underline{T} \otimes \underline{N})$

1 For the sake of simplicity, only the circular disk – also referred to as the penny-shaped crack – is considered.

(section 5.2). Throughout the chapter, the material surrounding the crack is linear elastic and isotropic[2].

5.1. Griffith circular (penny-shaped) crack in mode I

Let Oz denote the axis normal to the crack plane ($\underline{N} = \underline{e}_z$). The crack is herein described as a disk of radius a ($r < a$ and $z = 0$ in cylindrical coordinates), embedded in an infinite medium.

Let us consider the uniaxial stress state $\boldsymbol{\sigma}^\infty = q\underline{e}_z \otimes \underline{e}_z$ at infinity ($q > 0$). The crack surfaces are stress free. Due to the symmetry of the problem with respect to the crack plane $z = 0$, we focus on the half-space $z > 0$. As in Chapter 3, this problem is denoted by $\mathcal{P}_I$ and is defined by the following boundary conditions[3]:

$$\mathcal{P}_I : \quad \begin{array}{ll} z = 0 \text{ and } 0 < r < a : & \sigma_{rz} = \sigma_{\theta z} = \sigma_{zz} = 0 \\ |\underline{z}| \to \infty : & \boldsymbol{\sigma} \to \boldsymbol{\sigma}^\infty \end{array} \qquad [5.1]$$

Again, two auxiliary problems $\mathcal{P}'_I$ and $\mathcal{P}''_I$ are introduced. In $\mathcal{P}'_I$, the crack surfaces are subjected to the pressure q and the stress state at infinity vanishes:

$$\mathcal{P}'_I : \quad \begin{array}{ll} z = 0 \text{ and } r < a : & \sigma_{rz} = \sigma_{\theta z} = 0; \quad \sigma_{zz} = -q \\ |\underline{z}| \to \infty : & \boldsymbol{\sigma} \to 0 \end{array} \qquad [5.2]$$

In turn, $\mathcal{P}''_I$ corresponds to the uniform stress state $q\underline{e}_z \otimes \underline{e}_z$ in the whole space, with the following "boundary conditions":

$$\mathcal{P}''_I : \quad \begin{array}{ll} z = 0 \text{ and } r < a : & \sigma_{rz} = \sigma_{\theta z} = 0; \quad \sigma_{zz} = q \\ |\underline{z}| \to \infty : & \boldsymbol{\sigma} \to q\underline{e}_z \otimes \underline{e}_z \end{array} \qquad [5.3]$$

2 The case of point forces can be found in [KAC 03].

3 The symmetry boundary conditions on the plane $z = 0$ for $r > a$ prescribe the shear stress $\sigma_{rz} = \sigma_{\theta z} = 0$ and the normal component $\xi_z = 0$.

The superposition principle in linear elasticity states that $\mathcal{P}_I = \mathcal{P}_I' + \mathcal{P}_I''$. This implies that the displacement field solution of $\mathcal{P}_I$ is the sum of the solutions of $\mathcal{P}_I'$ and $\mathcal{P}_I''$, the latter being the following homogeneous transformation:

$$\underline{\xi} = \frac{q}{E}\big(- \nu r \underline{e}_r + z \underline{e}_z\big) \qquad [5.4]$$

5.1.1. *Solution of* $\mathcal{P}_I'$

We now focus on the displacement field solution of $\mathcal{P}_I'$. Following the same approach as in Chapter 3, it is sought in the form:

$$\underline{\xi} = z\frac{\partial}{\partial z}\underline{\operatorname{grad}}\,\phi - (3-4\nu)\phi_{,z}\underline{e}_z + (1-2\nu)\underline{\operatorname{grad}}\,\phi \qquad [5.5]$$

the components of which in cylindrical coordinates read:

$$\begin{aligned} \xi_r &= z\frac{\partial^2 \phi}{\partial r \partial z} + (1-2\nu)\frac{\partial \phi}{\partial r} \\ \xi_\theta &= \frac{z}{r}\frac{\partial^2 \phi}{\partial \theta \partial z} + \frac{1-2\nu}{r}\frac{\partial \phi}{\partial \theta} \\ \xi_z &= z\frac{\partial^2 \phi}{\partial z^2} - 2(1-\nu)\frac{\partial \phi}{\partial z} \end{aligned} \qquad [5.6]$$

where ϕ is a harmonic Neuber–Papkovitch potential. Using as before the notation $a_{,\alpha} = \partial a/\partial \alpha$, the corresponding components of the strain tensor are[4]:

$$\begin{aligned} \varepsilon_{rr} &= z\phi_{,rrz} + (1-2\nu)\phi_{,rr} \\ \varepsilon_{\theta\theta} &= -z\big(\phi_{,rrz} + \phi_{,zzz}\big) - (1-2\nu)\big(\phi_{,rr} + \phi_{,zz}\big) \\ \varepsilon_{zz} &= z\phi_{,zzz} - (1-2\nu)\phi_{,zz} \end{aligned} \qquad [5.7]$$

4 Recall that $\Delta\phi = 0$.

and

$$\begin{aligned} \varepsilon_{r\theta} &= \frac{1}{r}\big((1-2\nu)(\phi_{,r\theta} - \frac{1}{r}\phi_{,\theta}) + z(\phi_{,r\theta z} - \frac{1}{r}\phi_{,\theta z})\big) \\ \varepsilon_{\theta z} &= \frac{z}{r}\phi_{,\theta zz} \\ \varepsilon_{rz} &= z\phi_{,rzz} \end{aligned} \quad [5.8]$$

In turn, the components of the stress tensor read:

$$\begin{aligned} \sigma_{rr} &= 2\mu\big(z\phi_{,rrz} - 2\nu\phi_{,zz} + (1-2\nu)\phi_{,rr}\big) \\ \sigma_{\theta\theta} &= 2\mu\big(-(1-2\nu)\phi_{,rr} - \phi_{,zz} - z(\phi_{,rrz} + \phi_{,zzz})\big) \\ \sigma_{zz} &= 2\mu\big(z\phi_{,zzz} - \phi_{,zz}\big) \end{aligned} \quad [5.9]$$

and

$$\begin{aligned} \sigma_{r\theta} &= \frac{2\mu}{r}\big((1-2\nu)(\phi_{,r\theta} - \frac{1}{r}\phi_{,\theta}) + z(\phi_{,r\theta z} - \frac{1}{r}\phi_{,\theta z})\big) \\ \sigma_{\theta z} &= 2\mu\frac{z}{r}\phi_{,\theta zz} \\ \sigma_{rz} &= 2\mu z\phi_{,rzz} \end{aligned} \quad [5.10]$$

It appears that the boundary conditions [5.2] on the tangential stress components in the crack plane ($z = 0$) are automatically satisfied. The remaining conditions deal with the normal stress on the crack surface and with the normal displacement in the crack plane in the domain $r > a$. They are the counterparts of [3.40] and [3.41]:

$$z = 0 \text{ and } r < a: \quad \sigma_{zz} = -2\mu\phi_{,zz} = -q \quad [5.11]$$

$$z = 0 \text{ and } r > a: \quad \xi_z = -2(1-\nu)\phi_{,z} = 0 \quad [5.12]$$

We now seek a harmonic potential $\phi(r, \theta, z)$ meeting the conditions [5.11] and [5.12]. The cylindrical symmetry of the problem suggests that $\partial\phi/\partial\theta = 0$. We therefore consider its

Hankel transform $\mathcal{H}\phi(\eta, z)$ related to $\phi(r, z)$ as defined at [3.99] and [3.100]:

$$\phi(r, z) = \int_0^\infty \eta \mathcal{H}\phi(\eta, z) J_o(\eta r) d\eta \qquad [5.13]$$

Recalling Bessel's equation satisfied by the Bessel function J_o (see section 3.4), it is evident that the condition $\Delta\phi = 0$ can be reduced to the ordinary differential equation w.r.t. the variable z:

$$\frac{\partial^2 \mathcal{H}\phi}{\partial z^2} - \eta^2 \mathcal{H}\phi = 0 \qquad [5.14]$$

In the domain $z > 0$, the remote boundary condition $\boldsymbol{\sigma} \to 0$ as $z \to +\infty$ suggests that we consider the solution $\mathcal{A}(\eta)\exp(-\eta z)$ of [5.14], with which [5.13] reads:

$$\phi(r, z) = \int_0^\infty \eta \mathcal{A}(\eta) \exp(-\eta z) J_o(\eta r) d\eta \qquad [5.15]$$

Equation [5.11] and [5.12] now, respectively, take the form:

$$\begin{aligned} &(a)\ r < a : \int_0^\infty \eta A(\eta) J_o(\eta r) d\eta = \frac{q}{2\mu} \\ &(b)\ r > a : \int_0^\infty A(\eta) J_o(\eta r) d\eta = 0 \end{aligned} \qquad [5.16]$$

where $A(\eta) = \eta^2 \mathcal{A}(\eta)$. The second equation [5.16b] is satisfied by any function $A(\eta) = \sin(\eta\rho)$ provided that $\rho < a$ (see equation [3.98]). For this reason, we seek the function $A(\eta)$ in the integral form:

$$A(\eta) = \int_0^a g(\rho) \sin(\eta\rho)\, d\rho \qquad [5.17]$$

We are left with the determination of the unknown $g(\rho)$ meeting [5.16a]. This question has been considered in section 3.4. Indeed, the solution of [5.16a] is also the solution of [3.107] with $p(\rho) = q/2\mu$ and of the Abel's equation (see equation [3.103] and [3.110]):

$$0 < \rho < a : \quad g(\rho) = \frac{q\rho}{\pi\mu} \qquad [5.18]$$

The potential $\phi(r, z)$ of [5.15] eventually reads:

$$\phi(r, z) = \frac{q}{\pi\mu} \int_0^{\infty} \frac{1}{\eta} \Big(\int_0^a \rho \sin(\eta\rho)\, d\rho \Big) \exp(-\eta z) J_o(\eta r) d\eta \qquad [5.19]$$

This form can be used to derive the displacement (see equation [5.6]) and the stresses (see equations [5.9] and [5.10]). For illustrative purposes, let us derive the normal displacement ξ_z of the crack surfaces. From [5.6], in which we let $z = 0$, we obtain $\xi_z = -2(1 - \nu)\phi_{,z}$ with:

$$\phi_{,z}(r, z = 0) = -\frac{q}{\pi\mu} \int_0^{\infty} \Big(\int_0^a \rho \sin(\eta\rho)\, d\rho \Big) J_o(\eta r) d\eta, \qquad [5.20]$$

which also reads:

$$\phi_{,z}(r, z = 0) = -\frac{q}{\pi\mu} \int_0^a \rho \Big(\int_0^{\infty} \sin(\eta\rho) J_o(\eta r) d\eta \Big)\, d\rho \qquad [5.21]$$

Recalling [3.98] and the condition $r < a$, this retuns:

$$\phi_{,z}(r, z = 0) = -\frac{q}{\pi\mu} \int_r^a \frac{\rho}{\sqrt{\rho^2 - r^2}}\, d\rho = -\frac{q}{\pi\mu}\sqrt{a^2 - r^2} \qquad [5.22]$$

which in turn returns:

$$z = 0 : \quad \xi_z = \frac{4}{\pi} \frac{q(1 - \nu^2)}{E} \sqrt{a^2 - r^2} \qquad [5.23]$$

5.1.2. *Solution of* $\mathcal{P}_I$

As previously stated, the displacement solution ξ_z^I of problem $\mathcal{P}_I$ (boundary conditions [5.1]) is obtained from the sum of the solution ξ_z of $\mathcal{P}_I''$ (see [5.4]) and of [5.23] (solution of $\mathcal{P}_I'$). Since the solution ξ_z of $\mathcal{P}_I''$ vanishes for $z = 0$, we find that ξ_z^I is identical to [5.23] for $r < a$ and $z = 0^+$ (upper crack lip). The symmetry of the problem w.r.t. the crack plane $z = 0$ implies that $\xi_z(r, -z) = -\xi_z(r, z)$ and accordingly, $[\![\xi_z]\!] = 2\xi_z(r, z = 0^+)$. It follows that the normal displacement jump reads:

$$[\![\xi_z]\!] = \frac{8}{\pi} \frac{q(1-\nu^2)}{E} \sqrt{a^2 - r^2} \qquad [5.24]$$

The determination of the normal stress σ_{zz} in the crack plane ($z = 0$, $r > a$) requires that of the second derivative $\phi_{,zz}$ (see [5.9]):

$$\begin{aligned} \phi_{,zz}(r, z = 0) &= \frac{q}{\pi\mu} \int_0^\infty \Big(\int_0^a \rho\eta \sin(\eta\rho)\, d\rho \Big) J_o(\eta r) d\eta \\ &= -\frac{q}{\pi\mu} \int_0^\infty \Big(\int_0^a \rho \frac{\partial}{\partial\rho} \cos(\eta\rho)\, d\rho \Big) J_o(\eta r) d\eta \end{aligned} \qquad [5.25]$$

Integrating by part w.r.t. ρ in the second line of [5.25] yields:

$$\begin{aligned} \phi_{,zz}(r, z = 0) = \frac{q}{\pi\mu} \Big(&\int_0^a \Big(\int_0^\infty J_o(\eta r) \cos(\eta\rho)\, d\eta \Big) d\rho \\ &- a \int_0^\infty J_o(\eta r) \cos(\eta a)\, d\eta \Big) \end{aligned} \qquad [5.26]$$

It follows from [3.97] that[5]:

$$\begin{aligned} \phi_{,zz}(r, z = 0) &= \frac{q}{\pi\mu} \Big(\int_0^a \frac{d\rho}{\sqrt{r^2 - \rho^2}} - \frac{a}{\sqrt{r^2 - a^2}} \Big) \\ &\frac{q}{\pi\mu} \Big(\arcsin \frac{a}{r} - \frac{a}{\sqrt{r^2 - a^2}} \Big) \end{aligned} \qquad [5.27]$$

5 Recall that $r > a$.

Eventually, by applying the superposition principle ($\mathcal{P}_I = \mathcal{P}'_I + \mathcal{P}''_I$) and recalling that the stress field solution to $\mathcal{P}''_I$ is the uniform stress state $q\,\underline{e} \otimes \underline{e}_z$, we find from [5.9] and [5.27] that:

$$\sigma^I_{zz} = \frac{2}{\pi} q \big(\frac{a}{\sqrt{r^2 - a^2}} + \frac{\pi}{2} - \arcsin \frac{a}{r} \big) \qquad [5.28]$$

Defining the stress intensity factor in a similar fashion as in section 3.2 by $K_I = \lim_{r \to a^+} \sqrt{2(r-a)} \sigma^I_{zz}$, it is evident from [5.28] that:

$$K_I = \frac{2}{\pi} q \sqrt{a} \qquad [5.29]$$

Comparing [5.24] and [5.29] to [3.70] and [3.67], it is interesting to observe that the displacement jumps and the stress intensity factors derived in plane strain conditions and in axisymmetric conditions are identical up to a factor $2/\pi$.

5.2. Griffith circular (penny-shaped) crack under shear loading

In this section devoted to shear loading, it will be convenient to introduce a cartesian coordinate system (z_1, z_2, z_3) together with cylindrical coordinates $r = \sqrt{z_1^2 + z_2^2}$, θ and $z = z_3$. The boundary conditions are now given by:

$$\mathcal{P}_{II} : \quad \begin{array}{ll} z_3 = 0 \text{ and } r < a : & \sigma_{33} = \sigma_{32} = \sigma_{31} = 0 \\ |\underline{z}| \to \infty : & \boldsymbol{\sigma} \to \boldsymbol{\sigma}^{\infty} \end{array} \qquad [5.30]$$

in which the stress state at infinity reads:

$$\boldsymbol{\sigma}^{\infty} = q \left(\underline{e}_1 \otimes \underline{e}_3 + \underline{e}_3 \otimes \underline{e}_1 \right)$$

Clearly, with the boundary conditions [5.30], the solution of $\mathcal{P}_{II}$ is only defined up to a rigid body motion. In the following, we seek a displacement solution that is equivalent[6] to:

$$\underline{\xi}_{\infty}^{II} = \frac{q z_3}{\mu} \underline{e}_1 \qquad [5.31]$$

at infinity, that is, far from the pertubation caused by the crack.

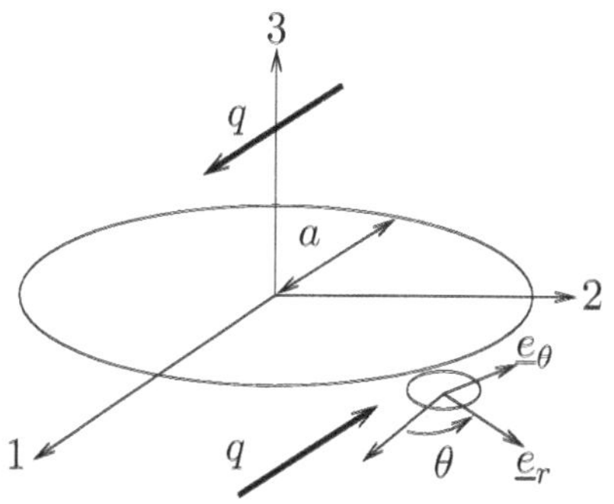

Figure 5.1. *Shear loading* $\sigma_{13}^{\infty} = q$

The idea is to determine the solution in the half-space $z_3 > 0$ first, and then in the half-space $z_3 < 0$ by appropriate symmetry conditions. Similarly to [3.71], we anticipate that they take the form[7]:

$$\begin{aligned} \xi_1^{II-}(z_1, z_2, z_3) &= -\xi_1^{II+}(z_1, z_2, -z_3) \\ \xi_2^{II-}(z_1, z_2, z_3) &= -\xi_2^{II+}(z_1, z_2, -z_3) \\ \xi_3^{II-}(z_1, z_2, z_3) &= \xi_3^{II+}(z_1, z_2, -z_3) \end{aligned} \qquad [5.32]$$

If $\underline{\xi}^{II+}$ is a solution of Navier equation in the domain $z_3 > 0$, this is also true of $\underline{\xi}^{II-}$ in the domain $z_3 < 0$. [5.32] in turn imply that $\xi_1^{II} = \xi_2^{II} = \sigma_{33}^{II} = 0$ on the plane $z_3 = 0$.

6 Two functions $f(x)$ and $g(x)$ are said to be equivalent at infinity if $f/g \to 1$ as $x \to \infty$.

7 Except on the upper and lower crack surfaces, that is for $z_3 = 0$ and $r < a$.

Again, two auxiliary problems $\mathcal{P}'_{II}$ and $\mathcal{P}''_{II}$ are introduced. Owing to [5.32], only the domain $z_3 > 0$ is considered. In $\mathcal{P}'_{II}$, the crack surfaces are subjected to the shear stress $-q$ and the stress state at infinity vanishes:

$$\mathcal{P}'_{II}: \quad \begin{array}{ll} z_3 = 0 \text{ and } r < a: & \sigma_{33} = \sigma_{32} = 0; \quad \sigma_{31} = -q \\ z_3 = 0 \text{ and } r > a: & \xi_1 = \xi_2 = \sigma_{33} = 0 \\ |\underline{z}| \to \infty: & \boldsymbol{\sigma} \to 0 \end{array} \qquad [5.33]$$

In turn, $\mathcal{P}''_{II}$ corresponds to the uniform stress state $\boldsymbol{\sigma} = \boldsymbol{\sigma}^\infty$ in the whole space, with the following “boundary conditions”:

$$\mathcal{P}''_{II}: \quad \begin{array}{ll} z_3 = 0 \text{ and } r < a: & \sigma_{33} = \sigma_{32} = 0; \quad \sigma_{31} = q \\ z_3 = 0 \text{ and } r > a: & \xi_1 = \xi_2 = \sigma_{33} = 0 \\ |\underline{z}| \to \infty: & \boldsymbol{\sigma} \to q\,(\underline{e}_1 \otimes \underline{e}_3 + \underline{e}_3 \otimes \underline{e}_1) \end{array} \qquad [5.34]$$

The displacement fields that are solutions to $\mathcal{P}'_{II}$ and $\mathcal{P}''_{II}$ are defined up to rigid body motions. Recalling that $\underline{\xi}^{II}$ is equivalent to $\underline{\xi}^{II}_\infty$ at infinity (see equation [5.31]), we therefore seek displacement fields that meet the condition $\xi_3 \to 0$ at infinity. It is then evident that the displacement field solution to [5.34] reads:

$$\underline{\xi} = \underline{\xi}^{II}_\infty = \frac{q}{\mu} z_3 \underline{e}_1 \qquad [5.35]$$

The superposition principle in linear elasticity states that $\mathcal{P}_{II} = \mathcal{P}'_{II} + \mathcal{P}''_{II}$. This motivates the following section.

5.2.1. *Solution of* $\mathcal{P}'_{II}$

We now focus on the displacement field solution to $\mathcal{P}'_{II}$. The latter is sought as a function of a harmonic potential $\phi(r, z_3)$ in the form [SEG 50]:

$$\underline{\xi} = \frac{1}{2\mu}\left(-4(1-\nu)\phi_{,3}\underline{e}_1 - 2(1-2\nu)\phi_{,1}\underline{e}_3 + 2z_3\frac{\partial}{\partial z_1}(\underline{\text{grad}}\,\phi)\right) \qquad [5.36]$$

In plane $z_3 = 0$, it is evident that ξ_2 and σ_{33} vanish, as required from [5.33], irrespective of the choice of $\phi(r, z_3)$. In this plane, the following expressions are derived:

$$\xi_1 = -\frac{2(1-\nu)}{\mu}\phi_{,3} \tag{5.37}$$

$$\xi_3 = -\frac{1-2\nu}{\mu}\phi_{,1} \tag{5.38}$$

$$\sigma_{31} = -2(1-\nu)\phi_{,33} + 2\nu\phi_{,11} \tag{5.39}$$

$$\sigma_{32} = 2\nu\phi_{,12} \tag{5.40}$$

$\phi(r, z_3)$ has the same expression as in [5.15]. With the same notation $A(\eta) = \eta^2\mathcal{A}(\eta)$ as in section 5.2, we find that the condition $\xi_1 = 0$ for $r > a$ takes the form [5.16b]. $A(\eta)$ is therefore sought in the form [5.17], so that $\phi(r, z_3)$ reads:

$$\phi(r, z_3) = \int_0^\infty \frac{1}{\eta}\Big(\int_0^a g(\rho)\sin(\eta\rho)\,d\rho\Big)\exp(-\eta z_3)J_o(\eta r)d\eta \tag{5.41}$$

Equation [5.39] together with [5.33] suggests that we aim to find the function $g(\rho)$ such that the second derivatives $\phi_{,11}$ and $\phi_{,33}$ are constant for $z_3 = 0$ and $r < a$. To begin with, we have:

$$\begin{aligned}\phi_{,33}(r, z_3 = 0) &= \int_0^\infty \Big(\int_0^a g(\rho)\eta\sin(\eta\rho)\,d\rho\Big)J_o(\eta r)d\eta \\ &= -\int_0^\infty \Big(\int_0^a g(\rho)\frac{\partial}{\partial\rho}\cos(\eta\rho)\,d\rho\Big)J_o(\eta r)d\eta\end{aligned} \tag{5.42}$$

Assuming that $g(0) = 0$, an integration by parts yields:

$$\phi_{,33}(r, z_3 = 0) = \int_0^a g'(\rho)\Big(\int_0^\infty J_o(\eta r)\cos(\eta\rho)\,d\eta\Big)d\rho \\ -g(a)\int_0^\infty J_o(\eta r)\cos(\eta a)\,d\eta \qquad [5.43]$$

Recalling [3.97], it is evident that the second integral vanishes[8]:

$$\phi_{,33}(r, z_3 = 0) = \int_0^r \frac{g'(\rho)}{\sqrt{r^2 - \rho^2}}d\rho \qquad [5.44]$$

For the above integral to be a constant, we choose $g(\rho) = \alpha\rho$ (see equation [3.104]), that is:

$$r < a: \quad \phi_{,33}(r, z_3 = 0) = \frac{\pi\alpha}{2} \qquad [5.45]$$

Observing that $r_{,1} = z_1/r$, it follows from [5.41] that:

$$\phi_{,1}(r, z_3 = 0) = \frac{\alpha z_1}{r}\int_0^\infty \Big(\int_0^a \rho\sin(\eta\rho)\,d\rho\Big)J_o'(\eta r)d\eta \qquad [5.46]$$

Observing that $\partial/\partial\eta\big(J_o(\eta r)/r\big) = J_o'(\eta r)$, an integration by parts w.r.t. η in [5.46] yields[9]:

$$\phi_{,1}(r, z_3 = 0) = -\frac{\alpha z_1}{r^2}\int_0^\infty \Big(\int_0^a \rho^2\cos(\eta\rho)\,d\rho\Big)J_o(\eta r)d\eta \qquad [5.47]$$

that is:

$$\phi_{,1}(r, z_3 = 0) = -\frac{\alpha z_1}{r^2}\int_0^a \rho^2\Big(\int_0^\infty \cos(\eta\rho)J_o(\eta r)d\eta\Big)\,d\rho \qquad [5.48]$$

---

8 $r < a$ and $\rho < a$.

9 Recall that $\lim_{x\to\infty} J_o(x) = 0$.

Using [3.97] with $r < a$ again, we eventually find that:

$$\phi_{,1}(r, z_3 = 0) = -\frac{\alpha z_1}{r^2} \int_0^r \frac{\rho^2}{\sqrt{r^2 - \rho^2}} \, d\rho = -\frac{\pi \alpha z_1}{4} \qquad [5.49]$$

It follows that the second derivative $\phi_{,11}(r, z_3 = 0)$ is a constant for $r < a$:

$$r < a: \quad \phi_{,11}(r, z_3 = 0) = -\frac{\pi \alpha}{4} \qquad [5.50]$$

Hence, the choice of linear function $g(\rho) = \alpha \rho$ in the definition [5.17] of $A(\eta)$ is compatible with the requirement that the shear stress σ_{31} be uniform. More precisely, the unknown coefficient α is determined from [5.39] together with [5.44] and [5.50]:

$$\alpha = \frac{2q}{\pi(2 - \nu)} \qquad [5.51]$$

Finally, we find from [5.40] together with [5.49] that the boundary condition [5.33] on σ_{32} is also satisfied. This proves that the displacement solution is indeed given by [5.36], where the potential $\phi(r, z_3)$ reads:

$$\phi(r, z_3) = \frac{2q}{\pi(2 - \nu)} \int_0^\infty \frac{e^{-\eta z_3}}{\eta} J_o(\eta r) \Big(\int_0^a \rho \sin(\eta \rho) \, d\rho \Big) d\eta \qquad [5.52]$$

We are left with the determination of the components ξ_1 and ξ_3 of the displacement of the upper crack surface. Starting from [5.37], the expression of ξ_1 involves the same integral as in [5.21]:

$$\xi_1(r, z_3 = 0) = \frac{4q(1 - \nu)}{\pi \mu (2 - \nu)} \int_0^a \rho \Big(\int_0^\infty J_o(\eta r) \sin(\eta \rho) \, d\eta \Big) d\rho \qquad [5.53]$$

that is (see equation [3.98]):

$$r < a \text{ and } z_3 = 0: \quad \xi_1 = \frac{4q}{\mu} \frac{1-\nu}{\pi(2-\nu)} \sqrt{a^2 - r^2} \tag{5.54}$$

In turn, combining [5.38] and [5.49] produces the following expression of ξ_3:

$$r < a \text{ and } z_3 = 0: \quad \xi_3 = \frac{q}{\mu} \frac{1-2\nu}{2(2-\nu)} z_1 \tag{5.55}$$

Using [3.97] in [5.48] for $r > a$ we find that:

$$\begin{aligned} r > a: \quad \phi_{,1}(r, z_3 = 0) &= -\frac{\alpha z_1}{r^2} \int_0^a \frac{\rho^2}{\sqrt{r^2 - \rho^2}} d\rho \\ &= -\alpha z_1 \int_0^{\frac{a}{r}} \frac{x^2}{\sqrt{1 - x^2}} dx \end{aligned} \tag{5.56}$$

that is:

$r > a:$

$$\phi_{,1}(r, z_3 = 0) = -\frac{\alpha z_1}{2} \left(\arctan \frac{a}{\sqrt{r^2 - a^2}} - \frac{a}{r} \sqrt{1 - \left(\frac{a}{r}\right)^2} \right) \tag{5.57}$$

Equations [5.38] and [5.51] then yield:

$r > a \text{ and } z_3 = 0:$

$$\xi_3 = \frac{q}{\mu} \frac{1-2\nu}{\pi(2-\nu)} z_1 \left(\arctan \frac{a}{\sqrt{r^2 - a^2}} - \frac{a}{r} \sqrt{1 - \left(\frac{a}{r}\right)^2} \right) \tag{5.58}$$

5.2.2. *Solution of* $\mathcal{P}_{II}$

We obtain the displacement field $\underline{\xi}^{II}$ solution to $\mathcal{P}_{II}$ (boundary conditions [5.30]) from the superposition of the fields of [5.54] and [5.55] (problem $\mathcal{P}'_{II}$) and of [5.35] (problem $\mathcal{P}''_{II}$). Since the solution of $\mathcal{P}''_{II}$ vanishes for $z_3 = 0$, we find that $\underline{\xi}^{II}$ is identical to the solution of $\mathcal{P}'_{II}$ (see equation [5.54] and [5.55]) in the crack plane. For $r < a$ and $z_3 = 0^+$ (upper crack lip).

$$r < a \text{ and } z_3 = 0^+ : \quad \begin{aligned} \xi_1^{II} &= \frac{4q}{\mu}\frac{1-\nu}{\pi(2-\nu)}\sqrt{a^2-r^2} \\ \xi_2^{II} &= 0 \\ \xi_3^{II} &= \frac{q}{\mu}\frac{1-2\nu}{2(2-\nu)}z_1 \end{aligned} \qquad [5.59]$$

In turn, for $r > a$ we have $\xi_1^{II} = \xi_2^{II} = 0$ and:

$$r > a \text{ and } z_3 = 0^+ :$$

$$\xi_3^{II} = \frac{q}{\mu}\frac{1-2\nu}{\pi(2-\nu)}z_1\left(\arctan\frac{a}{\sqrt{r^2-a^2}} - \frac{a}{r}\sqrt{1-\left(\frac{a}{r}\right)^2}\right) \qquad [5.60]$$

Recalling the symmetry conditions [5.32], the displacement jump reads:

$$[\![\underline{\xi}]\!] = 2\xi_I^{II+}\underline{e}_1 = \frac{q}{\mu}\frac{8(1-\nu)}{\pi(2-\nu)}\sqrt{a^2-r^2}\underline{e}_1 \qquad [5.61]$$

We conclude this section with the determination of the shear stress acting on the plane $z_3 = 0$ and discuss the issue of the stress intensity factor. Recalling [5.39] and [5.40], we first consider the expressions of the second derivatives $\phi_{,11}$,

$\phi_{,12}$ and $\phi_{,33}$ in the domain $r > a$ of the plane $z_3 = 0$. From [5.43], it is evident that:

$$r > a: \quad \phi_{,33}(r, z_3 = 0) = \alpha \left(\int_0^a \frac{d\rho}{\sqrt{r^2 - \rho^2}} - \frac{a}{\sqrt{r^2 - a^2}} \right) \qquad [5.62]$$

that is:

$$r > a: \phi_{,33}(r, z_3 = 0) = \alpha \left(\arctan \frac{a}{\sqrt{r^2 - a^2}} - \frac{a}{\sqrt{r^2 - a^2}} \right) \qquad [5.63]$$

In turn, $\phi_{,11}$ is best determined from [5.57]: Recalling that $r_{,1} = z_1/r$, we obtain:

$$r > a: \phi_{,11}(r, z_3 = 0) = \alpha \left(- \int_0^{\frac{a}{r}} \frac{x^2}{\sqrt{1 - x^2}} dx + \frac{a^3 z_1^2}{r^4 \sqrt{r^2 - a^2}} \right) \qquad [5.64]$$

which takes the form:

$$r > a: \quad \phi_{,11}(r, z_3 = 0) = \alpha \left(\frac{1}{2} \left(\frac{a}{r} \sqrt{1 - \left(\frac{a}{r} \right)^2} - \arctan \frac{a}{\sqrt{r^2 - a^2}} \right) + \frac{a^3 z_1^2}{r^4 \sqrt{r^2 - a^2}} \right) \qquad [5.65]$$

Again from [5.57], it is evident that:

$$r > a: \quad \phi_{,12}(r, z_3 = 0) = \alpha \frac{a^3 z_1 z_2}{r^4 \sqrt{r^2 - a^2}} \qquad [5.66]$$

Eventually, we apply the superposition principle ($\mathcal{P}_I = \mathcal{P}_I' + \mathcal{P}_I''$) while recalling that the stress field solution to $\mathcal{P}_I''$ is the uniform stress state $\boldsymbol{\sigma}^{\infty} = q\,(\underline{e}_1 \otimes \underline{e}_3 + \underline{e}_3 \otimes \underline{e}_1)$. From [5.39] and [5.40], we obtain:

$$
\begin{aligned}
r > a : \sigma_{13}(r, z_3 = 0) &= \sigma_{13}^{\infty} + 2\nu\phi_{,11}(r, z_3 = 0) \\
&\quad -2(1-\nu)\phi_{,33}(r, z_3 = 0) \\
r > a : \sigma_{23}(r, z_3 = 0) &= 2\nu\phi_{,12}(r, z_3 = 0)
\end{aligned}
\qquad [5.67]
$$

The shear stress acting on the plane $z_3 = 0$ is conveniently split into a radial component σ_{rz_3} (mode II) and an orthoradial component $\sigma_{\theta z_3}$ (mode III):

$$
\begin{aligned}
\sigma_{rz_3} &= \frac{z_1}{r}\sigma_{13} + \frac{z_2}{r}\sigma_{23} \\
\sigma_{\theta z_3} &= -\frac{z_2}{r}\sigma_{13} + \frac{z_1}{r}\sigma_{23}
\end{aligned}
\qquad [5.68]
$$

In turn, the stress intensity factors in mode II and mode III are defined as:

$$
\begin{aligned}
K_{II} &= \lim_{r \to a^+} \sqrt{2(r-a)}\,\sigma_{rz_3} \\
K_{III} &= \lim_{r \to a^+} \sqrt{2(r-a)}\,\sigma_{\theta z_3}
\end{aligned}
\qquad [5.69]
$$

From [5.63], [5.65] and [5.66], the followings limits are derived:

$$
\begin{aligned}
&\lim_{r \to a^+} \sqrt{2(r-a)}\,\phi_{,11} = \alpha \frac{z_1^2}{a\sqrt{a}} \\
&\lim_{r \to a^+} \sqrt{2(r-a)}\,\phi_{,12} = \alpha \frac{z_1 z_2}{a\sqrt{a}} \\
&\lim_{r \to a^+} \sqrt{2(r-a)}\,\phi_{,33} = -\alpha\sqrt{a}
\end{aligned}
\qquad [5.70]
$$

Introducing [5.67], [5.68] and [5.70] into [5.69], the stress intensity factors are found to be[10]:

$$
\begin{aligned}
K_{II}(\theta) &= \frac{4\cos\theta}{\pi(2-\nu)} q\sqrt{a} \\
K_{III}(\theta) &= -\frac{4(1-\nu)\sin\theta}{\pi(2-\nu)} q\sqrt{a}
\end{aligned}
\qquad [5.71]
$$

10 θ is the polar angle of the cylindrical coordinate system: $\cos\theta = z_1/r$ and $\sin\theta = z_2/r$.

6

Ellipsoidal Crack Model: the Eshelby Approach

Following the same idea as in section 4.3, an alternative to the three-dimensional (3D) Griffith crack model is to describe the initial configuration of the crack model, as a flat (oblate) ellipsoid with a symmetry around the Oz_3 axis (spheroid), centered at the origin of the space. The cross-section by the plane $z_3 = 0$ is a circle with radius a. The equation of the spheroid boundary reads:

$$\frac{z_1^2 + z_2^2}{a^2} + \frac{z_3^2}{c^2} = 1 \qquad [6.1]$$

with the aspect ratio $X = c/a \ll 1$. Again, the crack can be interpreted as an inhomogeneity of vanishing stiffness. This section briefly revisits the problem addressed in sections 5.1 and 5.2 in the light of the solution to Eshelby's inhomogeneity problem (section 2.4.4).

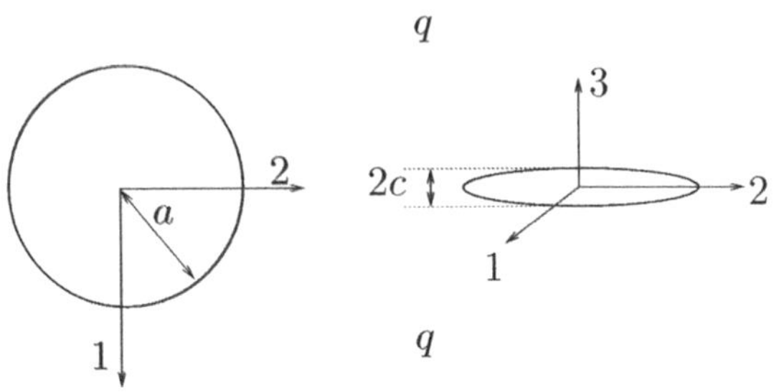

Figure 6.1. *The spheroidal crack model*

6.1. Mode I

We follow the line of reasoning presented in section 4.3. Mode I is defined by a uniaxial stress state at infinity:

$$\boldsymbol{\sigma}^{\infty} = q\,\underline{e}_3 \otimes \underline{e}_3 \qquad [6.2]$$

with $q > 0$. The corresponding strain state at infinity is derived from the linear elastic state equation:

$$\boldsymbol{\varepsilon}^{\infty} = \mathbb{C}^{-1} : \boldsymbol{\sigma}^{\infty} = \frac{q}{E}\left(\underline{e}_3 \otimes \underline{e}_3 - \nu\big(\underline{e}_1 \otimes \underline{e}_1 + \underline{e}_2 \otimes \underline{e}_2\big)\right) \qquad [6.3]$$

Section 2.4.4 shows that the ellipsoidal crack (inhomogeneity I) undergoes a homogeneous strain ε^I given by [2.70] in which **E** is replaced by the above value of ε^{∞} and $\mathbb{C}^I = 0$. The coefficients of $\mathbb{P}^I$ have been discussed in section 2.6.2. In the system of cylindrical coordinates with respect to the Oz_3 axis ($\underline{z} = r\underline{e}_r + z\underline{e}_3$, $z = z_3$, $\underline{e}_z = \underline{e}_3$), this produces:

$$\begin{aligned} \varepsilon^I_{rr} = \varepsilon^I_{\theta\theta} &= -\frac{q}{E}\left(\frac{(1-\nu)(1+2\nu)}{2} + \mathcal{O}(X)\right) \\ \varepsilon^I_{zz} &= \frac{q}{E}\left(\frac{4(1-\nu^2)}{\pi X} + \mathcal{O}(X^0)\right) \end{aligned} \qquad [6.4]$$

The displacement on the boundary of the ellipse is given by[1] $\underline{\xi}(\underline{z}) = \boldsymbol{\varepsilon}^I \cdot \underline{z}$, with $z = \pm X\sqrt{a^2 - r^2}$. The leading terms of its components read :

$$\begin{aligned} \xi_r &= -\frac{q}{2E}(1-\nu)(1+2\nu)r \\ \xi_\theta &= 0 \\ \xi_z &= \pm\frac{4q}{\pi E}(1-\nu^2)\sqrt{a^2-r^2} \end{aligned} \quad [6.5]$$

Since the ellipsoidal model assumes infinitesimal values of the aspect ratio X, the displacement jump across the upper and lower crack surfaces can be defined as in [4.110]:

$$[\![\underline{\xi}]\!](r) = \lim_{X\to 0}\left(\underline{\xi}(r, +X\sqrt{a^2-r^2}) - \underline{\xi}(r, -X\sqrt{a^2-r^2})\right) \quad [6.6]$$

that is:

$$[\![\underline{\xi}]\!](r) = 2\sqrt{a^2-r^2}\lim_{X\to 0}\left(X\boldsymbol{\varepsilon}^I\right)\cdot \underline{e}_z \quad [6.7]$$

This yields:

$$[\![\underline{\xi}]\!] = 8(1-\nu^2)\frac{q}{\pi E}\sqrt{a^2-r^2}\,\underline{e}_z \quad [6.8]$$

This derivation of the displacement jump confirms the result of the Griffith crack model (see equation [5.24]).

Since the aspect ratio X is infinitesimal by nature, the strain ε_{zz} is expected to take values beyond the domain of small pertubations. It can then be argued that it is questionable to apply the solution derived in section 2.4.4, since the latter is explicitly based on linear elasticity. More

1 In mode I, symmetry arguments show that the deformation gradient in the ellipsoidal inclusion has no skewsymmetric part.

precisely, the variation δX of the aspect ratio X is expected to be of the order of magnitude X. This means that a non-negligible geometry change takes place in the direction of the normal to the crack, so that the determination of the displacement ξ_z from $\varepsilon^I_{zz}z$ seems dubious. Note that similar criticisms can clearly be formulated concerning the developments based on the Griffith crack models.

A more satisfactory way to present the derivation of [6.8] is to adopt a rate type formulation. Mathematically speaking, it entails replacing the stress q by the rate $\dot{q}$ in [6.2]–[6.5]: The stresses, strains and displacements are, respectively, replaced by stress rates, strain rates and velocities. In this incremental reasoning, X is a variable depending on the loading q characterizing the aspect ratio of the current geometry. Hence, the current geometry of the crack boundary ∂I is defined by $z = \pm X(q)\sqrt{a^2 - r^2}$, where $X(q)$ is the value of the aspect ratio reached for the current loading q. The third equation of [6.4] is now replaced by[2]:

$$\dot{\varepsilon}^I_{zz} = \frac{\dot{q}}{E}\frac{4(1-\nu^2)}{\pi X(q)} \qquad [6.9]$$

At any point on ∂I, the normal component $\dot{\xi}_z$ of the velocity is determined as $\dot{\xi}_z = \dot{\varepsilon}^I_{zz}z$. An interesting mathematical feature is that the result only depends on $\dot{q}$, irrespective of the aspect ratio $X(q)$, since the latter vanishes in the product of $\dot{\varepsilon}^I_{zz}$ ($\propto 1/X$) and z ($\propto X$):

$$\dot{\xi}_z = \pm\frac{4\dot{q}}{\pi E}(1-\nu^2)\sqrt{a^2 - r^2} \qquad [6.10]$$

It is evident that [6.5] is obtained from [6.10] by integration with respect to q. Eventually, the rate type formulation thus retrieves the result stated in [6.8].

2 Only the leading term is considered.

We now focus on the stress state on the boundary ∂I of the ellipsoid, in the plane $z = 0$, where the unit normal vector is $\underline{e}_r$. The boundary conditions prescribe $\sigma_{rr} = \sigma_{r\theta} = \sigma_{rz} = 0$. Furthermore, we observe that the components $\varepsilon_{\theta\theta}$, ε_{zz} given by [6.4] and $\varepsilon_{\theta z} = 0$ of the strain tensor are continuous[3] across ∂I. This allows us to determine the unknown stresses $\sigma_{\theta\theta}$ and σ_{zz}, as well as the unknown strain ε_{rr} in the solid[4]:

$$\sigma_{\theta\theta} = \frac{4q\nu}{\pi X}$$
$$\sigma_{zz} = \frac{4q}{\pi X} \qquad [6.11]$$
$$\varepsilon_{rr} = -\frac{4q\nu(1+\nu)}{\pi E X}$$

As the ellipsoid tends toward the penny-shaped crack, that is as $X \to 0$, we note from [6.11] that the stress components $\sigma_{\theta\theta}$ and $\sigma_{zz} \to \infty$. This property is the counterpart in the framework of the 3D model of the singular behavior at $r = a$ already emphasized in the 3D Griffith crack model by the stress intensity factor K_I (see equation [5.29]).

6.2. Mode II

Mode II is again defined by a shear stress state at infinity:

$$\boldsymbol{\sigma}^{\infty} = q\left(\underline{e}_1 \otimes \underline{e}_3 + \underline{e}_3 \otimes \underline{e}_1\right) \qquad [6.12]$$

with $q > 0$. The corresponding strain state at infinity is derived from the linear elastic state equation:

$$\boldsymbol{\varepsilon}^{\infty} = \mathbb{C}^{-1} : \boldsymbol{\sigma}^{\infty} = \frac{q}{2\mu}\left(\underline{e}_1 \otimes \underline{e}_3 + \underline{e}_3 \otimes \underline{e}_1\right) \qquad [6.13]$$

3 $\underline{e}_\theta$ and $\underline{e}_z$ belong to the plane that is tangent to the surface ∂I.

4 Only the leading term ($\propto 1/X$) is considered.

From [2.70] together with [2.126], we readily obtain the strain in the ellipsoidal inclusion I in the form:

$$\varepsilon^I = \varepsilon^I_{13}\left(\underline{e}_1 \otimes \underline{e}_3 + \underline{e}_3 \otimes \underline{e}_1\right) \qquad [6.14]$$

with

$$\varepsilon^I_{13} = \frac{1-\nu}{(2-\nu)}\frac{2q}{\pi\mu X} \qquad [6.15]$$

As in section 4.3.2, devoted to the elliptic model in mode II, the idea is to determine the geometrical transformation of the crack from the property that the deformation gradient $\text{grad}\,\underline{\xi}$ is uniform over the later (see section 2.4). Direct use of [2.72] (with $\mathbf{E} = \varepsilon^\infty$) yields:

$$\xi_{1,3} = -2R_{1313}\tau^I; \quad \xi_{3,1} = -2R_{3113}\tau^I \qquad [6.16]$$

with

$$\tau^I = -\frac{1-\nu}{2-\nu}\frac{4q}{\pi X} \qquad [6.17]$$

Recalling [2.125] and [2.126], we obtain:

$$\xi_{1,3} = \frac{4q(1-\nu)}{\pi\mu(2-\nu)X} + \mathcal{O}(X^0); \quad \xi_{3,1} = \frac{q(1-2\nu)}{2\mu(2-\nu)} + \mathcal{O}(X) \qquad [6.18]$$

The displacement field within the ellipsoidal inclusion and on its boundary now reads[5]:

$$\underline{\xi} = \xi_{1,3} z_3 \underline{e}_1 + \xi_{3,1} z_1 \underline{e}_3 \qquad [6.19]$$

5 Note that $\xi_{1,1} = \xi_{3,3} = 0$ since $\varepsilon^I_{11} = \varepsilon^I_{33} = 0$.

Introducing $z_3 = \pm X\sqrt{a^2 - r^2}$ into [6.19], this equation is found to be perfectly consistent with the displacement field [5.59] obtained within the 3D Griffith crack model.

We conclude with some remarks concerning the stress state on the circle $r = a$ in the plane $z_3 = 0$ (boundary ∂I of the spheroidal crack cavity). As stated in the study of mode I (see section 6.1), the boundary conditions prescribe $\sigma_{rr} = \sigma_{r\theta} = \sigma_{rz_3} = 0$ and the components $\varepsilon_{\theta\theta}$, ε_{33} and $\varepsilon_{\theta z_3} = 0$ of the strain tensor are continuous across ∂I. As opposed to the Griffith crack model (see equation [5.69]), the shear stress component σ_{rz_3} has no singularity in $r = a$ since $\sigma_{rz_3} = 0$ is prescribed by the boundary conditions on the crack wall.

In contrast, the shear stress component $\sigma_{z_3\theta}$ is indeed singular in the vicinity of the crack wall. Indeed, due to the continuity of the strain coefficient $\varepsilon_{\theta z_3}$ across the ellipsoidal crack wall (in the midplane $z_3 = 0$), the value of the latter in the solid ($r \to a^+$) is determined from [6.14]:

$$\varepsilon_{\theta z_3} = \underline{e}_\theta \cdot \boldsymbol{\varepsilon}^I \cdot \underline{e}_3 = -\sin\theta \varepsilon_{13}^I \qquad [6.20]$$

with ε_{13}^I given in [6.15]. Since $\varepsilon_{13}^I \propto 1/X$, $\sigma_{z_3\theta} = 2\mu\varepsilon_{\theta z_3}^I \propto 1/X$ as well and therefore tends to ∞ as $X \to 0$.

7

Energy Release Rate and Conditions for Crack Propagation

The criterion for crack propagation is considered in this chapter in a thermodynamics framework (see also [BUI 06, RIC 78]). Following the classical reasoning, the driving force of crack propagation, namely the energy release rate, is identified from the Clausius–Duhem inequality, and the criterion involved is written as a threshold on the latter. Its consistency with a criterion considering a stress intensity factor is then discussed.

7.1. Driving force of crack propagation

For the sake of simplicity, let us consider the bidimensional case and a rectilinear crack whose length $\ell(t)$ is time dependent. The material system is denoted by $\Omega(\ell)$. The loading is defined by three scalar loading parameters $\mathcal{C}_\xi(t)$, $\mathcal{C}_F(t)$, $\mathcal{C}_T(t)$, which are given functions of time. It is formally denoted by $\mathcal{C}(t) = (\mathcal{C}_\xi(t), \mathcal{C}_F(t), \mathcal{C}_T(t))$. Accordingly, the density of mass forces is $\underline{F}(\underline{z}, t) = \mathcal{C}_F(t)\underline{F}^0(\underline{z})$ and the boundary conditions on the stress vector $\underline{T}$ and the displacement $\underline{\xi}$ read:

$$\partial\Omega^T : \underline{T}(\underline{z}, t) = \mathcal{C}_T(t)\underline{T}^0(\underline{z}); \ \partial\Omega^\xi : \underline{\xi}(\underline{z}, t) = \mathcal{C}_\xi(t)\underline{\xi}^0(\underline{z}) \qquad [7.1]$$

The boundary $\partial\Omega(\ell)$ is made up of $\partial\Omega_0 = \partial\Omega^T \cup \partial\Omega^\xi$ and of the crack lips $\mathcal{F}(\ell) = \mathcal{F}^+ \cup \mathcal{F}^-$. $\partial\Omega_0$ is time independent, while $\mathcal{F}(\ell)$ evolves with crack propagation. The crack lips are free of stress. The unit normal to $\mathcal{F}^-$ (respectively, $\mathcal{F}^+$) is denoted by $\underline{N}$ (respectively $-\underline{N}$). Unilateral contact is disregarded.

The purpose of this section is the derivation of the dissipated energy $\mathcal{D}$. Since the material behavior is linear elastic, the physical origin of the latter is the crack propagation. In other words, we anticipate that $\mathcal{D}$ will vanish simultaneously with $\dot{\ell}$. In this case, assuming an isothermal evolution, the Clausius–Duhem inequality takes the form:

$$\mathcal{D} = \mathcal{P}_e - \dot{W} \geq 0 \qquad [7.2]$$

$\mathcal{P}_e$ is the power of external forces acting on $\Omega(\ell)$

$$\mathcal{P}_e = \int_{\Omega(\ell)} \rho \underline{F} \cdot \dot{\underline{\xi}}\, dS + \int_{\partial\Omega} \underline{T} \cdot \dot{\underline{\xi}}\, ds \qquad [7.3]$$

More precisely, we have:

$$\mathcal{P}_e = \mathcal{C}_F(t) \int_\Omega \rho \underline{F}^0 \cdot \dot{\underline{\xi}}\, dS + \mathcal{C}_T(t) \int_{\partial\Omega^T} \underline{T}^0 \cdot \dot{\underline{\xi}}\, ds + \dot{\mathcal{C}}_\xi(t) \int_{\partial\Omega^\xi} \underline{T} \cdot \underline{\xi}^0\, ds \qquad [7.4]$$

In [7.2], $W(t)$ is the elastic energy stored in $\Omega(\ell)$ at time t. The current displacement field solution to the problem defined by the loading $\mathcal{C}(t)$ and the current geometry defined by $\ell(t)$ is $\underline{\xi}(t) = \underline{\xi}(\mathcal{C}(t), \ell(t))$. Hence, apart from the loading, the other input of the problem, which is time dependent, is the crack length $\ell(t)$. At any time, the elastic energy $W(t)$ stored in the system therefore reads:

$$W(t) = \int_{\Omega(\ell)} \frac{1}{2} \varepsilon(\mathcal{C}, \ell) : \mathbb{C} : \varepsilon(\mathcal{C}, \ell) \, dS \tag{7.5}$$

where $\varepsilon(\mathcal{C}, \ell)$ is the strain field associated with $\underline{\xi}(\mathcal{C}, \ell)$. The velocity $\dot{\underline{\xi}}$ that intervenes in the integrals of [7.4] can be split into the sum of two terms:

$$\dot{\underline{\xi}} = \frac{\partial \underline{\xi}}{\partial \mathcal{C}}_{|\ell} \dot{\mathcal{C}} + \frac{\partial \underline{\xi}}{\partial \ell}_{|\mathcal{C}} \dot{\ell} \tag{7.6}$$

Introducing [7.6] into [7.4] produces a decomposition of the power of external forces:

$$\mathcal{P}_e = \mathcal{P}_{e|\ell} + \mathcal{P}_{e|\mathcal{C}} \tag{7.7}$$

In [7.7], $\mathcal{P}_{e|\ell}$ is the power of external forces in conditions of fixed geometry (i.e. fixed crack length):

$$\mathcal{P}_{e|\ell} = \left(\mathcal{C}_F(t) \int_{\Omega} \rho \underline{F}^d \cdot \frac{\partial \underline{\xi}}{\partial \mathcal{C}} \, dS + \mathcal{C}_T(t) \int_{\partial \Omega^T} \underline{T}^d \cdot \frac{\partial \underline{\xi}}{\partial \mathcal{C}} \, ds \right) \dot{\mathcal{C}} + \left(\int_{\partial \Omega^\xi} \underline{T} \cdot \underline{\xi}^d \, ds \right) \dot{\mathcal{C}}_\xi \tag{7.8}$$

By means of the principle of virtual power, $\mathcal{P}_{e|\ell}$ can also be interpreted as the strain power:

$$\mathcal{P}_{e|\ell} = \int_{\Omega(\ell)} \boldsymbol{\sigma} : \dot{\boldsymbol{\varepsilon}} \, dS = \int_{\Omega(\ell)} \frac{1}{2} \boldsymbol{\varepsilon} : \mathbb{C} : \dot{\boldsymbol{\varepsilon}} \, dS \tag{7.9}$$

which in turn represents the derivative of the elastic energy in conditions of fixed geometry:

$$\mathcal{P}_{e|\ell} = \frac{\partial W}{\partial \mathcal{C}}_{|\ell} \dot{\mathcal{C}} \qquad [7.10]$$

The second contribution $\mathcal{P}_{e|\mathcal{C}}$ to $\mathcal{P}_e$ represents the power of external forces associated with crack propagation at constant loading:

$$\mathcal{P}_{e|\mathcal{C}} = \left(\mathcal{C}_F(t) \int_{\Omega(\ell)} \rho \underline{F}^0 \cdot \frac{\partial \underline{\xi}}{\partial \ell} \, dS + \mathcal{C}_T(t) \int_{\partial\Omega^T} \underline{T}^0 \cdot \frac{\partial \underline{\xi}}{\partial \ell} \, ds \right) \dot{\ell} \qquad [7.11]$$

At this stage, it is convenient to introduce the work of given external forces, denoted by $\Phi(\underline{\xi})$:

$$\Phi(\underline{\xi}) = \mathcal{C}_F(t) \int_{\Omega(\ell)} \rho \underline{F}^0 \cdot \underline{\xi} \, dS + \mathcal{C}_T(t) \int_{\partial\Omega^T} \underline{T}^0 \cdot \underline{\xi} \, ds \qquad [7.12]$$

It is evident that:

$$\mathcal{P}_{e|\mathcal{C}} = \frac{\partial \Phi}{\partial \ell}_{|\mathcal{C}} \dot{\ell} \qquad [7.13]$$

Eventually, introducing [7.10] and [7.13] into [7.2], the dissipation takes the form:

$$\mathcal{D} = \left(\frac{\partial W}{\partial \mathcal{C}}_{|\ell} \dot{\mathcal{C}} + \frac{\partial \Phi}{\partial \ell}_{|\mathcal{C}} \dot{\ell} \right) - \left(\frac{\partial W}{\partial \mathcal{C}}_{|\ell} \dot{\mathcal{C}} + \frac{\partial W}{\partial \ell}_{|\mathcal{C}} \dot{\ell} \right) = G(\mathcal{C}, \ell)\dot{\ell} \qquad [7.14]$$

where $G(\mathcal{C}, \ell)$ represents the energy release rate, defined as the opposite of the derivative of the potential energy $W - \Phi$ with respect to the crack length in conditions of fixed loading:

$$G(\mathcal{C}, \ell) = -\frac{\partial}{\partial \ell} (W - \Phi)_{|\mathcal{C}} \qquad [7.15]$$

Thus defined, the energy release rate $G(\mathcal{C},\ell)$ appears as the thermodynamic force responsible for propagation. It is the reason for which the criterion for crack propagation is written as a condition to be satisfied by $G(\mathcal{C},\ell)$. The simplest way is to define a threshold G_c such that:

$$\begin{array}{lll} G(\mathcal{C},\ell) < G_c & \Rightarrow & \dot{\ell} = 0 \\ G(\mathcal{C},\ell) = G_c & \Rightarrow & \dot{\ell} \geq 0 \end{array} \qquad [7.16]$$

G_c is the critical energy, considered a material constant for so-called perfectly brittle materials. The literature has extensively discussed the possibility that G_c varies as a function of the crack length (quasi-brittle materials). The R-curve concept is an attempt to capture this but will not be discussed further herein.

Let us assume that G_c is a constant and that the crack propagation condition is met for a given loading $\mathcal{C}$, which means that $G(\mathcal{C},\ell) = G_c$. This equation implicitly relates the current values of the crack length and of the loading parameter. Let $d\ell > 0$ denote a crack length increment. If $G(\mathcal{C},\ell + d\ell) < G(\mathcal{C},\ell)$, crack propagation will stop if the loading is kept constant: the propagation is said to be stable. Otherwise, that is if $G(\mathcal{C},\ell + d\ell) > G(\mathcal{C},\ell)$, propagation will be unstable. We conclude that the sign of the second derivative of the potential energy is a criterion for stable or unstable propagation:

$$\frac{\partial^2}{\partial \ell^2}(W - \Phi)_{|\mathcal{C}} > 0 \quad \Leftrightarrow \quad \text{stable propagation} \qquad [7.17]$$

7.2. Stress intensity factor and energy release rate

The purpose of this section is to show the consistency of the criterion of propagation [7.16] with a criterion based on the concept of material toughness of the form $K_I \leq K_{Ic}$. To do so, we will show that the energy release rate is a quadratic function of the stress intensity factors.

First, the elastic energy can be put in the form[1]:

$$W = \frac{1}{2}\left(\int_{\partial\Omega_0} \underline{T}\cdot\underline{\xi}\, ds + \int_{\Omega(\ell)} \rho\underline{F}(t)\cdot\underline{\xi}\, dS\right) \quad [7.18]$$

As a result of [7.12], the potential energy is

$$W - \Phi = \frac{1}{2}\int_{\partial\Omega^{\xi}} \underline{T}\cdot\underline{\xi}\, ds - \frac{1}{2}\int_{\partial\Omega^{T}} \underline{T}\cdot\underline{\xi}\, ds - \frac{1}{2}\int_{\Omega(\ell)} \rho\underline{F}(t)\cdot\underline{\xi}\, dS \quad [7.19]$$

The energy release rate is derived from [7.19] and [7.15], with particular attention to the fact that that the derivative with respect to ℓ is taken with the loading defined by $\mathcal{C}(t)$ kept constant:

$$G(\mathcal{C},\ell) = -\frac{1}{2}\int_{\partial\Omega^{\xi}} \frac{\partial\underline{T}}{\partial\ell}\cdot\underline{\xi}\, ds + \frac{1}{2}\int_{\partial\Omega^{T}} \underline{T}\cdot\frac{\partial\underline{\xi}}{\partial\ell}\, ds + \frac{1}{2}\int_{\Omega(\ell)} \rho\underline{F}(t)\cdot\frac{\partial\underline{\xi}}{\partial\ell}\, dS \quad [7.20]$$

Considering again that

$$\frac{\partial}{\partial\ell}\left(\underline{\xi}_{|\partial\Omega^{\xi}}\right) = 0; \quad \frac{\partial}{\partial\ell}\left(\underline{T}_{|\partial\Omega^{T}}\right) = 0$$

[7.20] can be put into the more symmetric form:

$$G(\mathcal{C},\ell) = \frac{1}{2}\left(\int_{\partial\Omega_0} \underline{T}\cdot\frac{\partial\underline{\xi}}{\partial\ell} - \frac{\partial\underline{T}}{\partial\ell}\cdot\underline{\xi} ds + \int_{\Omega(\ell)} \rho\underline{F}(t)\cdot\frac{\partial\underline{\xi}}{\partial\ell}\, dS\right) \quad [7.21]$$

1 In the bidimensional framework considered, ds denotes an elementary length on the boundary while dS is a surface element in the plane.

We shall label (1) (respectively, (2)) the stresses and displacements at crack length ℓ (respectively, $\ell + d\ell$), so that:

$$\frac{\partial \underline{\xi}}{\partial \ell} = \frac{\underline{\xi}^{(2)} - \underline{\xi}^{(1)}}{d\ell}; \quad \frac{\partial \underline{T}}{\partial \ell} = \frac{\underline{T}^{(2)} - \underline{T}^{(1)}}{d\ell}$$

With these notations, [7.21] reads:

$$2G(\mathcal{C}, \ell)d\ell = D_{12} - D_{21} \qquad [7.22]$$

with

$$\begin{aligned} D_{12} &= \int_{\partial\Omega_0} \underline{T}^{(1)} \cdot \underline{\xi}^{(2)}\, ds + \int_{\Omega(\ell)} \rho \underline{F}^{(1)} \cdot \underline{\xi}^{(2)}\, dS \\ D_{21} &= \int_{\partial\Omega_0} \underline{T}^{(2)} \cdot \underline{\xi}^{(1)}\, ds + \int_{\Omega(\ell)} \rho \underline{F}^{(2)} \cdot \underline{\xi}^{(1)}\, dS \end{aligned} \qquad [7.23]$$

In [7.23], $\underline{F}^{(1)} = \underline{F}^{(2)} = \underline{F}$ has been used since the loading is kept constant. It is useful to observe that the volume integrals in [7.23] are not modified if they are taken on $\Omega(\ell + d\ell)$ instead of $\Omega(\ell)$. It is also emphasized that the boundary $\partial\Omega(\ell + d\ell)$ comprises $\partial\Omega_0$ as well as the crack lips $\mathcal{F}^{\pm}(\ell + d\ell)$:

$$\partial\Omega(\ell + d\ell) = \partial\Omega_0 \cup \mathcal{F}^{+}(\ell + d\ell) \cup \mathcal{F}^{-}(\ell + d\ell)$$

Recalling that $\underline{T}^{(2)} = 0$ on the crack lips $\mathcal{F}^{\pm}(\ell + d\ell)$, the principle of virtual power yields:

$$\begin{aligned} D_{21} &= \int_{\partial\Omega_0} \underline{T}^{(2)} \cdot \underline{\xi}^{(1)}\, ds + \int_{\Omega(\ell+d\ell)} \rho \underline{F}^{(2)} \cdot \underline{\xi}^{(1)}\, dS \\ &= \int_{\Omega(\ell+d\ell)} \boldsymbol{\sigma}^{(2)} : \boldsymbol{\varepsilon}^{(1)}\, dS \end{aligned} \qquad [7.24]$$

We now deal with the group denoted by D_{12} in the right-hand side of [7.22]. The idea is again to apply the principle of virtual power to the domain $\Omega(\ell + d\ell)$. But $\boldsymbol{\sigma}^{(1)}$ is now considered the stress field and $\underline{\xi}^{(2)}$ the displacement field. Since $\boldsymbol{\sigma}^{(1)} \cdot \underline{N} \neq 0$ on the newly created crack lips $d\mathcal{F}^{\pm}$, there are two missing boundary terms in the expression of D_{12} that must be introduced in order to identify the complete expression of the virtual power:

$$D_{12} + \int_{d\mathcal{F}^-} \underline{T}^{(1)} \cdot \underline{\xi}^{(2)}\, ds + \int_{d\mathcal{F}^+} \underline{T}^{(1)} \cdot \underline{\xi}^{(2)}\, ds = \int_{\Omega(\ell+d\ell)} \boldsymbol{\sigma}^{(1)} : \boldsymbol{\varepsilon}^{(2)}\, dS \qquad [7.25]$$

in which $d\mathcal{F}^{\pm} = \mathcal{F}^{\pm}(\ell + d\ell) \setminus \mathcal{F}^{\pm}(\ell)$. As a result of the momentum balance condition $\operatorname{div} \boldsymbol{\sigma}^{(1)} = 0$, the stress vectors $\underline{T}^{(1)}$ acting on $d\mathcal{F}^-$ and $d\mathcal{F}^+$ are opposite:

$$\underline{T}^{(1)}_{|d\mathcal{F}^-} = \boldsymbol{\sigma}^{(1)} \cdot \underline{N} = -\underline{T}^{(1)}_{|d\mathcal{F}^+} \qquad [7.26]$$

Since the displacement field $\underline{\xi}^{(2)}$ undergoes a discontinuity across the crack, [7.25] reduces to:

$$D_{12} - \int_{d\mathcal{F}^-} [\![\underline{\xi}^{(2)}]\!] \cdot \boldsymbol{\sigma}^{(1)} \cdot \underline{N}\, ds = \int_{\Omega(\ell+d\ell)} \boldsymbol{\sigma}^{(1)} : \boldsymbol{\varepsilon}^{(2)}\, dS \qquad [7.27]$$

with $[\![\underline{\xi}^{(2)}]\!] = \underline{\xi}^{(2)}_{|d\mathcal{F}^+} - \underline{\xi}^{(2)}_{|d\mathcal{F}^-}$ Owing to the Maxwell–Betti reciprocity theorem, the right-hand sides in [7.24] and [7.27] are equal. Together with [7.22], this yields:

$$2G(\boldsymbol{C}, \ell)d\ell = \int_{d\mathcal{F}^-} [\![\underline{\xi}^{(2)}]\!] \cdot \boldsymbol{\sigma}^{(1)} \cdot \underline{N}\, ds \qquad [7.28]$$

We recall that the crack remains a straight line during propagation. For the derivation of $[\![\underline{\xi}^{(2)}]\!] \cdot \boldsymbol{\sigma}^{(1)} \cdot \underline{N}$ and its integration over $d\mathcal{F}^-$, it will prove useful to consider a system of cartesian coordinates. The origin is the crack tip at crack length ℓ. Oy is normal to the crack so that $\underline{N} = \underline{e}_y$, and the x-axis is oriented in the direction of propagation. Accordingly, $d\mathcal{F}^-$ is the segment $[0, d\ell]$ of the x-axis and we have:

$$[\![\underline{\xi}^{(2)}]\!] \cdot \boldsymbol{\sigma}^{(1)} \cdot \underline{N} = \sigma_{yy}^{(1)} [\![\xi_y^{(2)}]\!] + \sigma_{xy}^{(1)} [\![\xi_x^{(2)}]\!]$$

The expressions of the components of $\boldsymbol{\sigma}^{(1)}$ can be derived from [3.14] in which $r = 0$ corresponds to the crack tip of the crack $\mathcal{F}(\ell)$. In this system of polar coordinates, $d\mathcal{F}^-$ is the segment defined by $\theta = 0$ and $r \in [0, d\ell]$ so that $r = x$. Accordingly, we have:

$$\sigma_{yy}^{(1)} = \frac{K_I}{\sqrt{2\pi x}}; \quad \sigma_{xy}^{(1)} = \frac{K_{II}}{\sqrt{2\pi x}} \qquad [7.29]$$

In turn, the expressions of $[\![\xi_x^{(2)}]\!]$ and $[\![\xi_y^{(2)}]\!]$ were given[2] in [3.16]. In the latter, the origin of the polar coordinates is the crack tip of the crack $\mathcal{F}(\ell + d\ell)$ and the crack lips are defined by $\theta = \pm\pi$. In particular, the relation $r = d\ell - x$ should be used on $d\mathcal{F}^-$:

$$[\![\xi_x^{(2)}]\!] = \sqrt{\frac{d\ell - x}{2\pi}} \frac{4(1-\nu)K_{II}}{\mu};$$

$$[\![\xi_y^{(2)}]\!] = \sqrt{\frac{d\ell - x}{2\pi}} \frac{4(1-\nu)K_I}{\mu} \qquad [7.30]$$

Combining [7.28], [7.29] and [7.30], we obtain Irwin's formula[3]:

$$G(\mathcal{C}, \ell) = \frac{1-\nu^2}{E} \left(K_I^2 + K_{II}^2\right) \qquad [7.31]$$

---

2 The stress intensity factors in states (1) and (2) are different but the correction would introduce a second-order term w.r.t. $d\ell$.

3 Where we have used that $\int_0^a \sqrt{\frac{a-x}{x}} dx = \frac{a\pi}{2}$.

PART 2

Homogenization of Microcracked Materials

8

Fundamentals of Continuum Micromechanics

We hereafter propose a short review of the basic ideas of the homogenization theory in view of its application to microcracked materials. The interested reader is referred, for instance, to [DOR 05], [DOR 06], [MUR 87], [NEM 93] and [ZAO 02].

8.1. Scale separation

The basic idea of continuum micromechanics lies in the fact that the constitutive behavior of the considered material can be derived from the response of the so-called *representative elementary volume* (*r.e.v.*), regarded as a structure, to an appropriate loading. The typical length scale ℓ of the *r.e.v.*, denoted by Ω in the sequel, should be small enough as compared to the characteristic dimension L of the whole structure, so as to enable the use of the differential tools of continuum mechanics. In addition, ℓ should also be large enough as compared to the characteristic length, say δ, of the heterogeneities:

$$\delta \ll \ell \ll L \qquad [8.1]$$

This is necessary to ensure that Ω is indeed representative of the studied material. In the case of a microcracked material, the radius of the cracks is a good estimate of δ.

In certain coordinate systems, the location of the *r.e.v.* in the macroscopic structure with typical length L is given by a position vector denoted by $\underline{x}$. Accordingly, $\Omega(\underline{x})$ represents an *r.e.v.* located at point $\underline{x}$ of the macroscopic scale. A refined scale is necessary in order to reveal the heterogeneities with typical size δ of the microscopic scale. We denote by $\underline{z}$ the position vector in the corresponding coordinates system. From now on, the average over the *r.e.v.* $\Omega(\underline{x})$ of some physical quantity described at the microscopic scale by the field $a(\underline{z})$ is denoted by $\overline{a}(\underline{x})$ and defined as:

$$\overline{a}(\underline{x}) = \frac{1}{|\Omega(\underline{x})|} \int_{\Omega(\underline{x})} a(\underline{z})\, dV_z \qquad [8.2]$$

We have to mathematically translate the fact that $\Omega(\underline{x})$ undergoes a given strain state or stress state. At the macroscopic scale, the latter is characterized by strain and stress tensors $\mathbf{E}(\underline{x})$ and $\mathbf{\Sigma}(\underline{x})$, respectively. At the microscopic scale, these strain and stress states are more precisely described by the strain and stress fields $\boldsymbol{\varepsilon}(\underline{z})$ and $\boldsymbol{\sigma}(\underline{z})$. The fluctuations in space at the microscopic scale of the strain and stress fields are clearly the counterpart of the heterogeneity of the *r.e.v.*.

We now have to clarify the link between $\mathbf{E}(\underline{x})$ and $\boldsymbol{\varepsilon}(\underline{z})$ on the one hand, and between $\mathbf{\Sigma}(\underline{x})$ and $\boldsymbol{\sigma}(\underline{z})$ on the other hand. To answer this question, we need to define the loading to which the *r.e.v.* is subjected. In the following, the reasoning is performed over a given *r.e.v.*, where the macroscopic argument $\underline{x}$ is omitted for the sake of clarity.

With regard to the geometrical model adopted for cracks, two options were described in the previous chapters:

– the Griffith model in which the crack is considered as a segment in 2D, and as a surface in 3D. A displacement discontinuity takes place between the crack lips;

– the ellipsoidal (respectively, elliptic) model that considers the crack as a volume (respectively, a surface) in 3D (respectively, 2D).

We shall see that the second option is particularly suitable in view of applying the tools of micromechanics of heterogeneous materials with random microstructures, such as the so-called Eshelby-based schemes. This geometrical model is later referred to as an "inhomogeneity model". The principle of this approach is introduced in sections 8.2.1 to 8.2.4. Nevertheless, the first option (the Griffith crack one), which was implemented successfully early on in some pioneering works, dealt with the damage induced by microcracks (see, e.g. [KAC 93, AND 86, KRA 96, PEN 02]). The main ideas will be presented in section 8.3.

8.2. Inhomogeneity model for cracks

In this section, the cracks are modeled as 3D domains in which the mechanical properties are different from those of the bulk. Therefore, there is no discontinuity of the displacement field. This property is important in view of the derivation of the so-called average rules.

8.2.1. *Uniform strain boundary conditions*

Following [HIL 67] and [HAS 83], the so-called uniform strain boundary conditions consist of prescribing the displacement $\underline{\xi}$ on the boundary $\partial\Omega$:

$$(\forall \underline{z} \in \partial\Omega) \quad \underline{\xi}(\underline{z}) = \mathbf{E} \cdot \underline{z} \qquad [8.3]$$

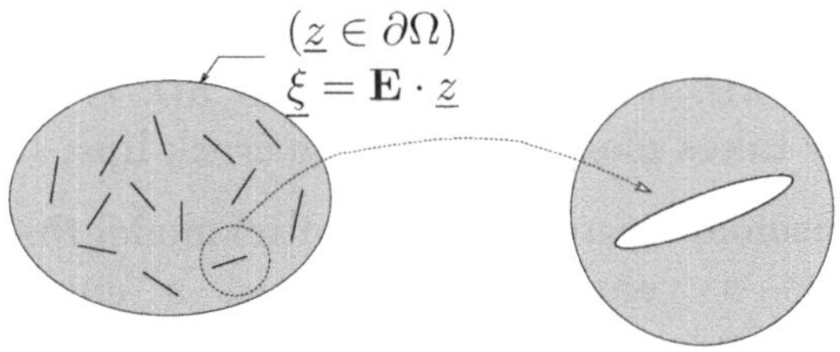

Figure 8.1. *Uniform strain conditions on the boundary of the r.e.v.*

This condition states that the displacement field on the boundary of the *r.e.v.* is identical to the displacement in the affine transformation in which the strain field is uniform and equal to E. In other words, irrespective of the local fluctuations of the strain field that take place *within* the *r.e.v.* when subjected to the loading defined by [8.3], the transformation that can be observed from the outside of the *r.e.v.* is identical to that of the affine transformation. Therefore, [8.3] mathematically translates the fact that Ω undergoes the strain state represented at the macroscopic scale by the tensor **E**.

We note, for further use, the rate-type formulation of the uniform strain boundary conditions:

$$(\forall \underline{z} \in \partial\Omega) \quad \dot{\underline{\xi}}(\underline{z}) = \dot{\mathbf{E}} \cdot \underline{z} \qquad [8.4]$$

For any displacement field $\underline{\xi}$ kinematically admissible with E in the sense of [8.3], the following "average rule" holds:

$$\overline{\nabla \underline{\xi}} = \mathbf{E} \qquad [8.5]$$

The proof of [8.5] is a straightforward consequence of the identity:

$$\int_{\partial\Omega} z_k n_j \, dS = \int_{\Omega} \frac{\partial z_k}{\partial z_j} \, dV = |\Omega| \delta_{jk} \qquad [8.6]$$

in which dS (respectively, dV) denotes an elementary surface (respectively, volume). Indeed, Green's formula together with [8.3] yields:

$$\int_{\Omega} \frac{\partial \xi_i}{\partial z_j} \, dV = \int_{\partial\Omega} \xi_i n_j \, dS_z = E_{ik} \int_{\partial\Omega} z_k n_j \, dS \qquad [8.7]$$

Combining [8.6] and [8.7], we obtain

$$\int_{\Omega} \frac{\partial \xi_i}{\partial z_j} \, dV = E_{ij} |\Omega| \qquad [8.8]$$

which is equivalent to the average rule [8.5]. Observing that $\mathbf{E}$ is a symmetric tensor, it is also found that the strain field ε meets the "strain average rule":

$$\overline{\varepsilon} = \frac{1}{2} (\overline{\nabla \underline{\xi}} + {}^t\overline{\nabla \underline{\xi}}) = \mathbf{E} \qquad [8.9]$$

Let us now consider an energy point of view. At the microscopic scale, the work of the internal forces associated with the stress field $\boldsymbol{\sigma}(\underline{z})$ in the strain rate $\dot{\varepsilon}$ is represented by the density $\boldsymbol{\sigma} : \dot{\varepsilon}$. In turn, at the macroscopic scale, the work of the internal forces in the *r.e.v.* is the product of the macroscopic stress tensor $\boldsymbol{\Sigma}$ and the macroscopic strain rate $\dot{\mathbf{E}}$. The energy consistency of the upscaling process requires that:

$$\int_{\Omega} \boldsymbol{\sigma} : \dot{\varepsilon} \, dV = \boldsymbol{\Sigma} : \dot{\mathbf{E}} |\Omega| \qquad [8.10]$$

In view of this discussion, let us first establish the identity

$$\int_{\partial\Omega} z_j \sigma_{ik} n_k \, dS = \overline{\sigma_{ij}} |\Omega| \qquad [8.11]$$

On the one hand, Green's formula yields

$$\int_{\Omega} \frac{\partial}{\partial z_k}(\sigma_{ik} z_j)\, dV = \int_{\partial\Omega} z_j \sigma_{ik} n_k\, dS \qquad [8.12]$$

On the other hand, owing to the momentum balance equation $\operatorname{div} \boldsymbol{\sigma} = 0$ at the microscopic scale, the left-hand side in [8.12] also reads:

$$\int_{\Omega} \frac{\partial}{\partial z_k}(\sigma_{ik} z_j)\, dV = \int_{\Omega} \delta_{jk} \sigma_{ik}\, dV = \overline{\sigma_{ij}} |\Omega| \qquad [8.13]$$

Equation [8.11] is obtained from the comparison of the two above equations. Let us now return to the discussion of energy consistency. The left-hand side in [8.10] is first transformed according to the principle of virtual work:

$$\int_{\Omega} \boldsymbol{\sigma} : \dot{\boldsymbol{\varepsilon}}\, dV = \int_{\partial\Omega} \dot{\xi}_i \sigma_{ik} n_k\, dS \qquad [8.14]$$

We then introduce the boundary condition [8.3] into the right-hand side of [8.14]:

$$\int_{\Omega} \boldsymbol{\sigma} : \dot{\boldsymbol{\varepsilon}}\, dV = \int_{\partial\Omega} z_j \dot{E}_{ij} \sigma_{ik} n_k\, dS \qquad [8.15]$$

The substitution of [8.11] into [8.15] now yields:

$$\int_{\Omega} \boldsymbol{\sigma} : \dot{\boldsymbol{\varepsilon}}\, dV = \overline{\boldsymbol{\sigma}} : \dot{\mathbf{E}} |\Omega| \qquad [8.16]$$

The above identity is referred to as Hill's Lemma. It reveals that the energy consistency condition summarized by [8.10] requires that the macroscopic stress tensor $\boldsymbol{\Sigma}$ be defined as the average $\overline{\boldsymbol{\sigma}}$ of the microscopic stress field:

$$\boldsymbol{\Sigma} = \overline{\boldsymbol{\sigma}} \qquad [8.17]$$

Note that the validity of Hill's lemma does not require that the stress and strain involved in [8.16] be related by any constitutive behavior. The only assumptions are that the velocity field $\dot{\underline{\xi}}$ associated with the strain field $\dot{\varepsilon}$ is kinematically admissible with $\dot{\mathbf{E}}$ in the sense of [8.4] and that $\operatorname{div}\boldsymbol{\sigma} = 0$. Moreover, there is obviously no need for the kinematic field to have the physical dimension of a velocity. Hence, we consider a displacement field $\underline{\xi}$ that is kinematically admissible with the macroscopic tensor $\mathbf{E}$ in the sense of [8.9] and the associated strain field $\boldsymbol{\varepsilon} = \frac{1}{2}(\nabla\underline{\xi} + {}^t\nabla\underline{\xi})$. If $\operatorname{div}\boldsymbol{\sigma} = 0$, then Hill's lemma reads:

$$\int_{\Omega} \boldsymbol{\sigma} : \boldsymbol{\varepsilon}\, dV = \overline{\boldsymbol{\sigma}} : \mathbf{E}|\Omega| \tag{8.18}$$

In the framework of a given constitutive behavior at the microscopic scale, the boundary conditions [8.3] (or [8.4]) together with the "stress average rule" define a boundary value problem which provides, at least theoretically, a methodology for determining the macroscopic stress–strain relation, that is the link between the macroscopic strain tensor $\mathbf{E}$ and the macroscopic stress tensor $\boldsymbol{\Sigma}$.

8.2.2. *Uniform stress boundary conditions*

In order to define a boundary value problem on the *r.e.v.*, an alternative possibility consists in prescribing the stress vector acting on the boundary $\partial\Omega$ in the following manner:

$$(\forall \underline{z} \in \partial\Omega) \quad \boldsymbol{\sigma}(\underline{z}) \cdot \underline{n}(\underline{z}) = \boldsymbol{\Sigma} \cdot \underline{n}(\underline{z}) \tag{8.19}$$

Equation [8.19] aims at translating the fact that the *r.e.v.* Ω undergoes the stress state represented at the macroscopic scale by $\boldsymbol{\Sigma}$. Introducing [8.19] together with [8.6] into [8.11] yields the stress average rule:

$$\overline{\boldsymbol{\sigma}} = \boldsymbol{\Sigma} \tag{8.20}$$

Observing that [8.14] still holds, we then introduce [8.19] into the latter:

$$\int_{\Omega} \sigma : \dot{\varepsilon} \, dV = \Sigma : \dot{\bar{\varepsilon}} |\Omega| \qquad [8.21]$$

Energy consistency [8.10] now requires that the macroscopic strain tensor be defined as the average of the microscopic strain field:

$$\mathbf{E} = \overline{\varepsilon} \qquad [8.22]$$

8.2.3. *Linear elasticity with uniform strain boundary conditions*

Let us illustrate the methodology of homogenization in the case of a microcracked linear elastic solid. The constitutive equation at point $\underline{z}$ of the microscopic scale then takes the linear form $\sigma = \mathbb{C}(\underline{z}) : \varepsilon$, where $\mathbb{C}(\underline{z})$ represents the local stiffness tensor. Indeed, a set of *opened* microcracks $\mathcal{C}_i$ may be represented as the limit case of a linear elastic material with vanishing stiffness. This means that $\mathbb{C}(\underline{z})$ is either equal to the stiffness tensor $\mathbb{C}^s$ of the solid domain[1] Ω^s or to 0 in the cracks:

$$\mathbb{C}(\underline{z}) = \begin{cases} \mathbb{C}^s & (\underline{z} \in \Omega^s) \\ \mathbb{C}_i^c = 0 & (\underline{z} \in \mathcal{C}_i) \end{cases} \qquad [8.23]$$

In the case of closed cracks, it will prove useful to introduce a non-zero fictitious stiffness tensor $\mathbb{C}_i^c$ for the crack $\mathcal{C}_i$ in order to take into account a non-zero stress state applied to the crack faces:

$$\mathbb{C}(\underline{z}) = \begin{cases} \mathbb{C}^s \ (\underline{z} \in \Omega^s) \\ \mathbb{C}_i^c \ (\underline{z} \in \mathcal{C}_i) \end{cases} \qquad [8.24]$$

1 For simplicity, the solid phase is assumed to be homogeneous.

The case of closed cracks is addressed in detail in section 10.4. At first sight, the assumption of linear elastic behavior suggests that the response of the *r.e.v.* to the loading defined by [8.3] linearly depends on $\mathbf{E}$. This leads to introduce the so-called strain concentration tensor $\mathbb{A}(\underline{z})$ relating the macroscopic and microscopic strain tensors:

$$\varepsilon(\underline{z}) = \mathbb{A}(\underline{z}) : \mathbf{E} \qquad [8.25]$$

The symmetrized unit tensor of the fourth order being denoted by $\mathbb{I}$, note that the strain average rule [8.9] implies that $\overline{\mathbb{A}} = \mathbb{I}$.

Introducing the strain concentration rule [8.25] into the microscopic state equation $\boldsymbol{\sigma} = \mathbb{C}(\underline{z}) : \varepsilon$ and recalling the stress average rule [8.17] yields the macroscopic state equation in the form of a linear relation between the macroscopic stress and strain tensors:

$$\boldsymbol{\Sigma} = \mathbb{C}^{\text{hom}} : \mathbf{E} \quad \text{with} \quad \mathbb{C}^{\text{hom}} = \overline{\mathbb{C} : \mathbb{A}} \qquad [8.26]$$

$\mathbb{C}^{\text{hom}}$ is referred to as a homogenized or effective stiffness tensor. Introducing the crack porosity φ, that is, the volume fraction of the cracks, and the volume fraction φ_i of the crack $\mathcal{C}_i$ ($\sum_i \varphi_i = \varphi$), [8.24] yields:

$$\begin{aligned}\mathbb{C}^{\text{hom}} &= (1-\varphi)\mathbb{C}^s : \overline{\mathbb{A}}^s + \sum_i \varphi_i \mathbb{C}_i^c : \overline{\mathbb{A}}^i \\ &= \mathbb{C}^s + \sum_i \varphi_i (\mathbb{C}_i^c - \mathbb{C}^s) : \overline{\mathbb{A}}^i \end{aligned} \qquad [8.27]$$

with

$$\overline{\mathbb{A}}^s = \frac{1}{|\Omega^s|} \int_{\Omega^s} \mathbb{A}(\underline{z})\, dV_z; \; \overline{\mathbb{A}}^i = \frac{1}{|\mathcal{C}_i|} \int_{\mathcal{C}_i} \mathbb{A}(\underline{z})\, dV_z \qquad [8.28]$$

Owing to [8.23], the effective stiffness in the case of opened cracks reduces to:

$$\mathbb{C}^{\text{hom}} = (1-\varphi)\mathbb{C}^s : \overline{\mathbb{A}}^s = \mathbb{C}^s : \big(\mathbb{I} - \sum_i \varphi_i \overline{\mathbb{A}}^i\big) \qquad [8.29]$$

However, despite the assumption of linear elasticity, a more careful analysis reveals that nonlinearity arises in the response of the *r.e.v.*, the origin of which is geometrical in nature (see also section 6.1). Indeed, the relative displacement of the crack lips may be of the order of the crack opening itself. This is readily seen, for instance in mode I, in the framework of an elliptic (respectively, ellipsoidal) model for the crack shape. Considering the expressions [4.109] (model of elliptic crack) and [6.5] (model of ellipsoidal crack) of the displacement component normal to the crack, it suffices that the ratio[2] q/E be of the order of the aspect ratio X for the normal component ξ_2 (respectively, ξ_z) to be of the order of the (initial) crack opening.

In order to overcome this difficulty, the loading of the *r.e.v.* should be defined on its current configuration from the macroscopic strain rate $\dot{\mathbf{E}}$ instead of $\mathbf{E}$, while prescribing the velocity $\dot{\underline{\xi}}$ on $\partial\Omega$ instead of the displacement. [8.3] is therefore replaced by [8.4]. The strain concentration rule [8.25] becomes a strain *rate* concentration rule:

$$\dot{\varepsilon}(\underline{z}) = \mathbb{A}(\underline{z}) : \dot{\mathbf{E}} \qquad [8.30]$$

while [8.26] is replaced by a state equation in the rate form:

$$\dot{\boldsymbol{\Sigma}} = \overline{\dot{\boldsymbol{\sigma}}} = \mathbb{C}^{\text{hom}} : \dot{\mathbf{E}} \quad \text{with} \quad \mathbb{C}^{\text{hom}} = \overline{\mathbb{C} : \mathbb{A}} \qquad [8.31]$$

2 The latter can be interpreted as the order of magnitude of the strain at infinity.

This means that the tensor $\mathbb{C}^{\text{hom}}$ in the above equation must *a priori* be interpreted as the *tangent* stiffness tensor of the equivalent homogeneous medium. Note that the strain rate concentration tensor $\mathbb{A}(\underline{z})$ in [8.31] now depends on the current configuration of the *r.e.v.*

8.2.4. *Linear elasticity with uniform stress boundary conditions*

When starting from the uniform stress boundary conditions [8.19], it is more convenient to introduce a microscopic constitutive equation at point $\underline{z}$ in the form $\dot{\varepsilon} = \mathbb{S}(\underline{z}):\dot{\sigma}$, where $\mathbb{S}(\underline{z})$ represents the local compliance tensor. By definition, the latter is the inverse of the local stiffness tensor $\mathbb{C}(\underline{z})$ introduced in section 8.2.3.

In the case of opened cracks (see equation [8.23]), this *a priori* leads to an infinite value of $\mathbb{S}(\underline{z})$. This mathematical difficulty will be avoided in temporarily considering a finite value $\mathbb{S}^c$ for the compliance of an opened crack. In practice, all quantities involving $\mathbb{S}^c$ are then implicitely replaced by their limit when $\mathbb{S}^c \to \infty$:

$$\mathbb{S}(\underline{z}) = \begin{cases} \mathbb{S}^s = \mathbb{C}^{s-1} & (\underline{z} \in \Omega^s) \\ \mathbb{S}^c \to \infty & (\underline{z} \in \mathcal{C}_i) \end{cases} \qquad [8.32]$$

Similarly, in the case of closed cracks, [8.24] readily yields:

$$\mathbb{S}(\underline{z}) = \begin{cases} \mathbb{S}^s = \mathbb{C}^{s-1} & (\underline{z} \in \Omega^s) \\ \mathbb{S}_i^c = \mathbb{C}_i^{c-1} & (\underline{z} \in \mathcal{C}_i) \end{cases} \qquad [8.33]$$

The loading parameter being now the macroscopic stress tensor Σ, the assumption of linearity leads to introduce the so-called stress (rate) concentration tensor $\mathbb{B}(\underline{z})$ relating the rates of the macroscopic and microscopic stress tensors:

$$\dot{\boldsymbol{\sigma}}(\underline{z}) = \mathbb{B}(\underline{z}) : \dot{\boldsymbol{\Sigma}} \qquad [8.34]$$

The stress average rule [8.20] implies that $\overline{\mathbb{B}} = \mathbb{I}$.

Introducing the stress concentration rule [8.34] into the microscopic state equation $\dot{\varepsilon} = \mathbb{S}(\underline{z}) : \dot{\boldsymbol{\sigma}}$, and recalling the strain average rule [8.22], yields the macroscopic state equation in the form of a linear relation between the macroscopic stress and strain rate tensors:

$$\dot{\mathbf{E}} = \mathbb{S}^{\text{hom}} : \dot{\boldsymbol{\Sigma}} \quad \text{with} \quad \mathbb{S}^{\text{hom}} = \overline{\mathbb{S} : \mathbb{B}} \qquad [8.35]$$

$\mathbb{S}^{\text{hom}}$ is referred to as a (tangent) homogenized compliance tensor. The homogenized compliance tensor $\mathbb{S}^{\text{hom}}$ thus derived and the homogenized stiffness tensor $\mathbb{C}^{\text{hom}}$ introduced in [8.31] are the inverse of one another under the assumption of equivalence of the uniform stress and uniform strain boundary conditions [HIL 67, MAN 72]. By analogy with [8.27], the consistency rule $\overline{\mathbb{B}} = \mathbb{I}$ yields:

$$\mathbb{S}^{hom} = \mathbb{S}^s + \sum_i \varphi_i (\mathbb{S}_i^c - \mathbb{S}^s) : \overline{\mathbb{B}}^i \qquad [8.36]$$

In the case of opened cracks, the product $\mathbb{S}_i^c : \overline{\mathbb{B}}^i$ in [8.36] is, at this stage, undetermined since $\mathbb{S}_i^c \to \infty$ and $\overline{\mathbb{B}}^i \to 0$. If we examine [8.27] and [8.36], it appears that the derivation of estimates of the effective stiffness or compliance of the microcracked material does not require a complete knowledge of the strain rate (stress rate) concentration tensor $\mathbb{A}(\underline{z})$ (respectively, $\mathbb{B}(\underline{z})$) over the whole *r.e.v.*: we observe that the knowledge of the average quantities $\overline{\mathbb{A}}^i$ (respectively, $\overline{\mathbb{B}}^i$) is in fact sufficient. Chapters 10 and 11 develop various methods for estimating the latter.

8.3. General results on homogenization with Griffith cracks

8.3.1. *Hill's lemma with Griffith cracks*

When moving from the inhomogeneity crack model to the Griffith crack model, a fundamental mathematical difference arises. It lies in the fact that the displacement field in the latter model is discontinuous across the crack lips. This implies that some derivations in sections 8.2.1 and 8.2.2 are no longer valid.

For each crack $\mathcal{C}_K$ in the *r.e.v.*, we introduce the upper crack lip $\mathcal{C}_K^+$ and the lower crack lip $\mathcal{C}_K^-$. The unit normal to $\mathcal{C}_K$ oriented from $\mathcal{C}_K^-$ toward $\mathcal{C}_K^+$ is denoted by $\underline{N}^K$. The displacement field on $\mathcal{C}_K^+$ (respectively, $\mathcal{C}_K^-$) is denoted by $\underline{\xi}^+$ (respectively, $\underline{\xi}^-$). The displacement jump reads $[\![\underline{\xi}]\!] = \underline{\xi}^+ - \underline{\xi}^-$.

Irrespective of the type of boundary conditions (uniform strain or stress), let us start with the left-hand side of [8.7]. The integration by part now includes a contribution of each crack lip:

$$\int_\Omega \frac{\partial \xi_i}{\partial z_j} dV = \int_{\partial\Omega} \xi_i n_j \, dS - \sum_K \int_{\mathcal{C}_K^+} \xi_i^+ N_j^K dS$$
$$+ \sum_K \int_{\mathcal{C}_K^-} \xi_i^- N_j^K dS \qquad [8.37]$$

which also reads[3]:

$$\int_\Omega \frac{\partial \xi_i}{\partial z_j} dV + \sum_K \int_{\mathcal{C}_K} [\![\xi_i]\!] N_j^K dS_z = \int_{\partial\Omega} \xi_i n_j \, dS_z \qquad [8.38]$$

For the same reason, the Hill lemma [8.16] has to be modified in order to account for the displacement

3 In [8.38], there is no need to differentiate $\mathcal{C}_K^\pm$ from $\mathcal{C}_K$.

discontinuities. Owing to the momentum balance equation, we have:

$$\int_\Omega \sigma_{ij}\frac{\partial \xi_i}{\partial z_j}\,dV = \int_{\partial\Omega} \sigma_{ij}\xi_i n_j\,dS - \sum_K \int_{C_K^+} \sigma_{ij}^+\xi_i^+ N_j^K dS$$

$$+\sum_K \int_{C_K^-} \sigma_{ij}^-\xi_i^- N_j^K dS \qquad [8.39]$$

where $\sigma_{ij}^\pm$ denote the values taken by the stress components σ_{ij} on each lip of the crack. As opposed to the displacement, the stress vector $\boldsymbol{\sigma} \cdot \underline{N}^K$ is continuous across the crack. This implies that $\sigma_{ij}N_j^K$ is continuous as well. [8.39] therefore also reads:

$$\int_\Omega \sigma_{ij}\frac{\partial \xi_i}{\partial z_j}\,dV + \sum_K \int_{C_K} \sigma_{ij}N_j^K [\![\xi_i]\!] dS = \int_{\partial\Omega} \sigma_{ij}\xi_i n_j\,dS \qquad [8.40]$$

The left-hand side can be physically interpreted as the total work of internal forces in the *r.e.v.*: it includes the contribution of the bulk (volume integral over Ω) and that of the stress vector acting on each crack lip.

8.3.2. *Uniform strain boundary conditions*

Let us now consider uniform strain boundary conditions. As defined in [8.3], [8.38] yields:

$$\int_\Omega \frac{\partial \xi_i}{\partial z_j}\,dV + \sum_K \int_{C_K} [\![\xi_i]\!] N_j^K dS_z = E_{ik}\int_{\partial\Omega} z_k n_j\,dS_z \qquad [8.41]$$

which replaces [8.7]. Using [8.6], we obtain:

$$\int_\Omega \nabla\underline{\xi}\,dV + \sum_K \int_{C_K} [\![\underline{\xi}]\!] \otimes \underline{N}^K\,dS = \mathbf{E}|\Omega| \qquad [8.42]$$

The symmetric part of the above equation in turn reads[4]:

$$\mathbf{E} = \overline{\varepsilon} + \frac{1}{|\Omega|} \sum_K \int_{\mathcal{C}_K} [\![\underline{\xi}]\!] \overset{s}{\otimes} \underline{N}^K \, dS \qquad [8.43]$$

Equation [8.43] states the generalized average rule in the presence of displacement discontinuities.

In a similar manner, we introduce [8.3] into [8.40]:

$$\int_\Omega \sigma_{ij} \frac{\partial \xi_i}{\partial z_j} \, dV + \sum_K \int_{\mathcal{C}_K} \sigma_{ij} N_j^K [\![\xi_i]\!] dS = E_{ik} \int_{\partial\Omega} z_k \sigma_{ij} n_j \, dS \quad [8.44]$$

Using [8.11][5], it appears that

$$\int_\Omega \boldsymbol{\sigma} : \boldsymbol{\varepsilon} \, dV + \sum_K \int_{\mathcal{C}_K} [\![\underline{\xi}]\!] \cdot \boldsymbol{\sigma} \cdot \underline{N}^K dS = \overline{\boldsymbol{\sigma}} : \mathbf{E} |\Omega| \qquad [8.45]$$

As already mentioned, the left-hand side can be physically interpreted as the total work of internal forces in the *r.e.v.*. The energy consistency of the homogenization process therefore requires that

$$\int_\Omega \boldsymbol{\sigma} : \boldsymbol{\varepsilon} \, dV + \sum_K \int_{\mathcal{C}_K} [\![\underline{\xi}]\!] \cdot \boldsymbol{\sigma} \cdot \underline{N}^K dS = \boldsymbol{\Sigma} : \mathbf{E} |\Omega| \qquad [8.46]$$

Comparing [8.45] and [8.46] reveals that $\boldsymbol{\Sigma} = \overline{\boldsymbol{\sigma}}$, which means that the stress average rule derived in [8.17] in the case of continuous displacement fields is not modified in the presence of displacement discontinuities.

---

4 $\mathbf{E}$ is assumed to be a symmetric tensor. $\underline{a} \overset{s}{\otimes} \underline{b}$ is the symmetric part of $\underline{a} \otimes \underline{b}$.

5 Note that the validity of [8.11] is clearly due to the fact that the momentum balance implies the continuity of the stress vector and of its components $\sigma_{ij} N_j$ across the crack.

8.3.3. *Uniform stress boundary conditions*

We now consider boundary conditions in the form [8.19]. The same reasoning as in section 8.2.2 (devoted to the inhomogeneity model) immediately yields the average rule [8.20].

We now introduce the boundary conditions [8.19] into [8.40]. We successively obtain:

$$\int_{\Omega} \sigma_{ij} \frac{\partial \xi_i}{\partial z_j} dV + \sum_K \int_{\mathcal{C}_K} \sigma_{ij} N_j [\![\xi_i]\!] dS = \Sigma_{ij} \int_{\partial\Omega} \xi_i n_j \, dS \qquad [8.47]$$

$$\int_{\Omega} \boldsymbol{\sigma} : \boldsymbol{\varepsilon} \, dV + \sum_K \int_{\mathcal{C}_K} [\![\underline{\xi}]\!] \cdot \boldsymbol{\sigma} \cdot \underline{N} dS = \boldsymbol{\Sigma} : \int_{\Omega} \underline{\xi} \otimes \underline{n} \, dS \qquad [8.48]$$

Since the left-hand side in [8.48] represents the work of the internal forces in the *r.e.v.*, the energy consistency of the homogenization process requires us to define the macroscopic strain as:

$$\mathbf{E} = \frac{1}{|\Omega|} \int_{\partial\Omega} \underline{\xi} \overset{s}{\otimes} \underline{n} \, dS \qquad [8.49]$$

Using [8.38], we formally retrieve the average rule [8.43] derived in the context of uniform strain boundary conditions:

$$\mathbf{E} = \overline{\boldsymbol{\varepsilon}} + \frac{1}{|\Omega|} \sum_K \int_{\mathcal{C}_K} [\![\underline{\xi}]\!] \overset{s}{\otimes} \underline{N}^K \, dS \qquad [8.50]$$

8.3.4. *Derivation of effective properties in linear elasticity: principle of the approach*

In sections 8.2.3 and 8.2.4, we presented the principle of the derivation of estimates of the effective stiffness or compliance tensors in the framework of the inhomogeneity model for cracks. We now deal with this issue for Griffith cracks. The *r.e.v.* is assumed to be made up of a homogeneous linear elastic material with a stiffness (respectively, compliance) tensor $\mathbb{C}^s$ (respectively, $\mathbb{S}^s$).

In sections 8.3.2 and 8.3.3, we have established that micro- and macroscopic stresses and strains are, respectively, related by $\boldsymbol{\Sigma} = \overline{\boldsymbol{\sigma}}$ and [8.43]. A combination of these "average rules" together with the local state equation $\boldsymbol{\varepsilon} = \mathbb{S}^s : \boldsymbol{\sigma}$ yields (see, e.g. [KAC 93] and [HOR 83]):

$$\mathbf{E} = \mathbb{S}^s : \boldsymbol{\Sigma} + \frac{1}{|\Omega|} \sum_k \int_{\mathcal{C}_k} [\![\underline{\xi}]\!] \overset{s}{\otimes} \underline{N}\, dS \quad [8.51]$$

The determination of the macroscopic state equation, which is the link between the macroscopic strain and stress tensors $\mathbf{E}$ and $\boldsymbol{\Sigma}$, thus requires an estimate of the displacement vector discontinuity $[\![\underline{\xi}]\!]$. In the case of a single Griffith crack embedded in an infinite homogeneous medium, the developments in Chapters 3 (plane strain conditions) and 5 (3D conditions) have proved that the displacement discontinuity linearly depends on the macroscopic stress at infinity. Returning to the case of a *r.e.v.*, this suggests to postulate a linear relationship between the displacement discontinuities and the macroscopic loading parameter for the boundary conditions at stake:

$$\forall \underline{z} \in \mathcal{C}_K : \quad [\![\underline{\xi}]\!](\underline{z}) = \boldsymbol{\alpha}^K(\underline{z}) : \mathbf{E} \quad [8.52]$$

or

$$\forall \underline{z} \in \mathcal{C}_K : \quad [\![\underline{\xi}]\!](\underline{z}) = \boldsymbol{\beta}^K(\underline{z}) : \boldsymbol{\Sigma} \quad [8.53]$$

where α^K and β^K are third-order concentration tensors. Introducing [8.52] or [8.53] into [8.51] yields the macroscopic state equation, at least theoretically.

In fact, in the case of a single crack embedded in an infinite continuum, we have already stated in Chapters 3 and 5 that a displacement jump across the crack arises only if the asymptotic stress state meets the condition $\sigma^\infty \cdot \underline{N} \neq 0$. It seems quite natural to assume that this statement remains valid in the case of multiple cracks. Accordingly, consider a given crack $\mathcal{C}_K$ with unit normal $\underline{N}^K$ located in the *r.e.v.*, the displacement jump is expected to depend linearly on the "stress vector" $\boldsymbol{\Sigma} \cdot \underline{N}^K$:

$$\forall \underline{z} \in \mathcal{C}_K : \quad [\![\underline{\xi}]\!] = \mathbf{a}^K(\underline{z}) \cdot \left(\boldsymbol{\Sigma} \cdot \underline{N}^K\right) \tag{8.54}$$

Using the symmetry of the Cauchy stress tensor $\boldsymbol{\Sigma}$, it is readily seen (see the Appendix, section 8.3.5) that

$$[\![\underline{\xi}]\!] \overset{s}{\otimes} \underline{N}^K = \mathbb{S}^K : \boldsymbol{\Sigma} \tag{8.55}$$

with

$$\mathbb{S}^K = \frac{1}{2}\left(\mathbf{a}^K \overline{\underline{\otimes}} \left(\underline{N}^K \otimes \underline{N}^K\right) + \left(\underline{N}^K \otimes \underline{N}^K\right) \overline{\underline{\otimes}}\, \mathbf{a}^K\right) \tag{8.56}$$

$\mathbb{S}^K$ represents a stress-based concentration tensor relating the local displacement jump to the macroscopic stress. $\mathbb{S}^K$ clearly possesses the minor symmetry. In order to establish that $\mathbb{S}^K$ also possesses the major symmetry, it is necessary to prove that $\mathbf{a}^K$ is itself a symmetric tensor. This result is readily verified in the case of an isotropic solid. It will be proved in the anisotropic case in a 2D framework (see equations [9.33] and [9.34]). Introducing [8.55] into [8.51] eventually yields the expression of the effective compliance $\mathbb{S}^{\text{hom}}$:

$$\mathbf{E} = \mathbb{S}^{\text{hom}} : \boldsymbol{\Sigma} \quad \text{with} \quad \mathbb{S}^{\text{hom}} = \mathbb{S}^s + \sum_K \overline{\mathbb{S}}^K \qquad [8.57]$$

with

$$\overline{\mathbb{S}}^K = \frac{1}{2}\left(\overline{\mathrm{a}}^K \overline{\underline{\otimes}} \left(\underline{N}^K \otimes \underline{N}^K\right) + \left(\underline{N}^K \otimes \underline{N}^K\right) \overline{\underline{\otimes}}\, \overline{\mathrm{a}}^K\right),$$

$$\overline{\mathrm{a}}^K = \frac{1}{|\Omega|} \int_{\mathcal{C}_K} \mathrm{a}^K dS \qquad [8.58]$$

$\overline{\mathbb{S}}^K$ physically represents the contribution of the crack $\mathcal{C}_K$ to the effective compliance. The above estimate of the effective compliance tensor can be interestingly compared to [8.35] and [8.36]. Both the inhomogeneity model and the Griffith crack model yield an expression of the effective compliance in which each crack introduces an additional correcting term to the compliance of the solid phase. At this stage, however, a difference exists between the two models: the compliance tensor of [8.36] relates the macroscopic strain and stress *rates* and as such is a tangent compliance tensor while [8.57] represents an elastic compliance tensor in the usual sense. The consistency of the two models will be proved in Chapter 10 (section 10.3).

Alternatively, if uniform strain boundary conditions are used, a fourth-order strain-based concentration tensor $\tilde{\mathbb{A}}^K$ is introduced in order to express that the displacement jump across crack $\mathcal{C}_K$ depends linearly on the macroscopic strain $\mathbf{E}$:

$$[\![\underline{\xi}]\!] \overset{s}{\otimes} \underline{N}^K = \tilde{\mathbb{A}}^K : \mathbf{E} \qquad [8.59]$$

After integration of [8.59] over the crack and introduction into [8.51], the following expression of the effective stiffness is derived:

$$\mathbb{C}^{\text{hom}} = \mathbb{C}^s : \left(\mathbb{I} - \sum_K \bar{\tilde{\mathbb{A}}}^K\right) \qquad [8.60]$$

which should be compared to [8.29].

Equations [8.57] and [8.60] provide two approaches of the effective elastic properties. They are in fact perfectly consistent under the assumption that uniform stress boundary conditions are equivalent to uniform strain boundary conditions. Indeed, the displacement jump vector is determined either from [8.54] (as a function of $\mathbf{\Sigma}$) or from [8.59] (as a function of $\mathbf{E}$). If both determinations are equivalent, we have:

$$[\![\underline{\xi}]\!] \overset{s}{\otimes} \underline{N}^K = \tilde{\mathbb{A}}^K : \mathbf{E} = \mathbb{S}^K : \mathbf{\Sigma} \qquad [8.61]$$

which yields $\tilde{\mathbb{A}}^K = \mathbb{S}^K : \mathbb{S}^{\text{hom}}$. With this expression of $\tilde{\mathbb{A}}^K$, [8.57] is readily retrieved from [8.60].

8.3.5. *Appendix*

The purpose of this section is to establish [8.55] and [8.56]. For the sake of simplicity, the index K of the crack is removed. Because of the symmetry of $\mathbf{\Sigma}$, we have:

$$a_{ik}\Sigma_{kl}N_l = a_{ik}\Sigma_{lk}N_l = a_{il}\Sigma_{kl}N_k \qquad [8.62]$$

From [8.54], we then successively obtain:

$$[\![\xi_i]\!] = \frac{1}{2}\left(a_{ik}N_l + a_{il}N_k\right)\Sigma_{kl} \qquad [8.63]$$

$$\begin{aligned}[\![\xi_i]\!]N_j + [\![\xi_j]\!]N_i = &\left(\frac{1}{2}\left(a_{ik}N_jN_l + a_{il}N_jN_k\right)\right. \\ &\left. +\frac{1}{2}\left(a_{jk}N_iN_l + a_{jl}N_iN_k\right)\right)\Sigma_{kl}\end{aligned} \qquad [8.64]$$

Considering two second-order tensors a and b, the tensorial product $\overline{\underline{\otimes}}$ is now defined as:

$$(\mathbf{a}\overline{\underline{\otimes}}\mathbf{b})_{ijkl} = \frac{1}{2}\left(a_{ik}b_{jl} + a_{il}b_{jk}\right) \qquad [8.65]$$

[8.64] then reads:

$$[\![\xi_i]\!]N_j + [\![\xi_j]\!]N_i = \left((\mathbf{a}\overline{\underline{\otimes}}(\mathbf{N}\otimes\mathbf{N}))_{ijkl} + ((\mathbf{N}\otimes\mathbf{N})\overline{\underline{\otimes}}\mathbf{a})_{ijkl}\right)\Sigma_{kl} \qquad [8.66]$$

[8.55] and [8.56] are the intrinsic formulation of the above equation.

9

Homogenization of Materials Containing Griffith Cracks

This chapter addresses the homogenization problem within the framework of the Griffith crack model. It is based on the displacement discontinuities derived in Chapters 3 (2D) and 5 (3D).

9.1. Dilute estimates in isotropic conditions

If Ω is a bounded domain in which the location of the cracks are characterized in a deterministic way, it is clear that the tensorial fields a^K and $\mathbb{S}^K$ of [8.54]–[8.56] and the corresponding macroscopic state equation will depend on the particular geometry of the considered *r.e.v.* We restrict ourselves to a simpler situation generally referred to as the dilute case. The idea is that the crack density being small enough, it seems reasonable to estimate the concentration tensors at a given crack as if the latter would be embedded in an infinite medium with stiffness $\mathbb{C}^s$. In addition, instead of the boundary conditions [8.3] or [8.19] on the boundary of Ω, we consider remote boundary conditions written at infinity.

The response of a single Griffith crack was already determined in mode I and II ([3.70] and [3.96] for plane strain conditions, and [5.24] and [5.61] in 3D conditions) within the framework of an isotropic solid. For illustrative purposes, we now apply the latter results to the determination of 3D isotropic effective properties. We therefore consider an isotropic distribution of crack orientations in an isotropic homogeneous solid[1], all cracks being described according to the circular model (see Chapter 5). Assuming a continuous distribution of crack orientations, we transform the discrete sum[2], which appears in [8.51] into an integral:

$$\mathbf{E} = \mathbb{S}^s : \mathbf{\Sigma} + \frac{\mathcal{N}}{4\pi} \int \sin\theta \, d\theta d\varphi \Big(\int_{\mathcal{C}_{\theta,\varphi}} [\![\underline{\xi}]\!] \overset{s}{\otimes} \underline{e}_r(\theta,\varphi) \, dS \Big) \quad [9.1]$$

where $\mathcal{N}$ denotes the crack density, that is the number of cracks per unit volume. The unit normal $\underline{N}$ to the crack $\mathcal{C}_{\theta,\varphi}$ is the unit radial vector $\underline{e}_r(\theta,\varphi)$ of the system of spherical coordinates with angles θ and φ.

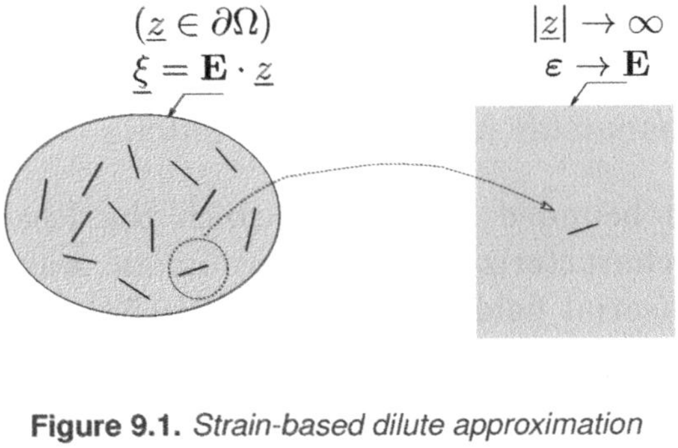

Figure 9.1. *Strain-based dilute approximation*

1 Bulk and shear moduli k^s, μ^s, Poisson coefficient ν^s, Young modulus E^s.

2 With respect to the index k on the cracks set.

9.1.1. *Stress-based dilute estimate of stiffness*

The assumption of isotropic effective behavior amounts to:

$$\mathbf{E} = \mathbb{S}^{\text{hom}} : \mathbf{\Sigma} \quad \text{with} \quad \mathbb{S}^{\text{hom}} = \frac{1}{3k^{\text{hom}}}\mathbb{J} + \frac{1}{2\mu^{\text{hom}}}\mathbb{K} \qquad [9.2]$$

In order to determine k^{hom} and μ^{hom}, we just need the particular form of the state equation for two independent stress states, typically a purely spherical one and a purely deviatoric one.

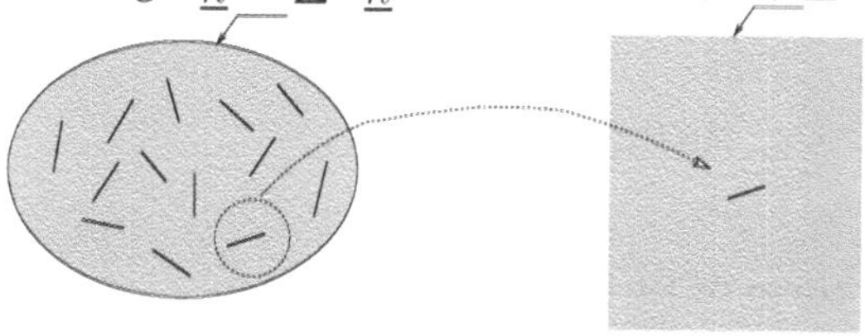

Figure 9.2. *Stress-based dilute approximation*

The principle of the stress-based dilute approximation is depicted in Figure 9.2 in which the right-hand side is used in order to derive an estimate of the tensor $\overline{\mathbb{S}}^K$ of [8.57]. To begin with, we consider the spherical stress state $\mathbf{\Sigma} = \Sigma\mathbf{1}$. Section 5.1 has determined the response of a crack with normal $\underline{e}_z$ to a remote uniaxial stress state $\boldsymbol{\sigma}^\infty = q\underline{e}_z \otimes \underline{e}_z$. As already observed (see introductions to Chapters 3 and 5), a remote biaxial stress (stresses in crack plane)

$$q(\mathbf{1} - \underline{e}_z \otimes \underline{e}_z)$$

does not induce any displacement jump. We therefore conclude that the displacement jump induced by the remote stress state

$\Sigma\mathbf{1}$ in a circular crack with normal $\underline{e}_r$ is identical to the one induced by the remote stress state $q\underline{e}_r \otimes \underline{e}_r$. It is therefore given by [5.24] in which q is replaced by Σ and $\underline{e}_z$ by $\underline{e}_r$:

$$\int_{\mathcal{C}_{\theta,\varphi}} [\![\underline{\xi}]\!] \overset{s}{\otimes} \underline{e}_r(\theta,\varphi)\, dS = \frac{16\Sigma(1-\nu^{s2})}{E^s}\Big(\int_0^a r\sqrt{a^2-r^2}\, rdr\Big)\underline{e}_r \quad [9.3]$$

that is

$$\int_{\mathcal{C}_{\theta,\varphi}} [\![\underline{\xi}]\!] \overset{s}{\otimes} \underline{e}_r(\theta,\varphi)\, dS = \frac{16a^3\Sigma(1-\nu^{s2})}{3E^s}\underline{e}_r \otimes \underline{e}_r \quad [9.4]$$

Noting that

$$\frac{1}{4\pi}\int \underline{e}_r \otimes \underline{e}_r \, \sin\theta\, d\theta\, d\varphi = \frac{1}{3}\mathbf{1} \quad [9.5]$$

the integral term of [9.1] reduces to:

$$\frac{\mathcal{N}}{4\pi}\int \sin\theta\, d\theta d\varphi\Big(\int_{\mathcal{C}_{\theta,\varphi}} [\![\underline{\xi}]\!] \overset{s}{\otimes} \underline{e}_r(\theta,\varphi)\, dS\Big) = \frac{16\mathcal{N}a^3\Sigma(1-\nu^{s2})}{9E^s}\mathbf{1} \quad [9.6]$$

We finally introduce [9.6] into [9.1]

$$\mathbf{E} = \Sigma\left(\frac{1}{3k^s} + \frac{16\mathcal{N}a^3(1-\nu^{s2})}{9E^s}\right)\mathbf{1} \quad [9.7]$$

which yields

$$k^{hom} = \frac{k^s}{1+Q_k d} \quad \text{with} \quad Q_k = \frac{16(1-\nu^{s2})}{9(1-2\nu^s)} \quad [9.8]$$

where d, defined as $\mathcal{N}a^3$, appears as a measurement of the effect of cracks on the bulk modulus. It is generally referred to as the crack density parameter [BUD 76].

As a second macroscopic stress state, we consider the deviatoric tensor

$$\mathbf{\Sigma} = \Sigma(\underline{e}_x \otimes \underline{e}_x - \underline{e}_z \otimes \underline{e}_z) \qquad [9.9]$$

The question is to determine the (vectorial) displacement jump induced in a crack, in which orientation is defined by the angles θ and φ. For a given orientation, this stress state will *a priori* induce both a normal and a tangential displacement jump. The normal component ξ_r of $[\![\underline{\xi}]\!]$ is related to the component Σ_{rr}. In turn, both $\Sigma_{r\theta}$ (respectively, $\Sigma_{r\varphi}$) will yield a contribution ξ_θ (respectively, ξ_φ) to the tangential part of the displacement jump. Accordingly, ξ_r is derived from [5.24] with q and $\underline{e}_z$ being replaced by Σ_{rr} and $\underline{e}_r$ respectively. Similarly, ξ_θ (respectively ξ_φ) is derived from [5.61] with q and $\underline{e}_1$ being replaced by $\Sigma_{r\theta}$ (respectively, $\Sigma_{r\varphi}$) and $\underline{e}_\theta$ (respectively, $\underline{e}_\varphi$). Eventually, the displacement jump reads[3]:

$$[\![\underline{\xi}]\!] = \frac{8(1-\nu^{s2})}{\pi E^s}\sqrt{a^2-r^2}\Big(\Sigma_{rr}\underline{e}_r + \frac{2}{2-\nu^s}(\Sigma_{r\theta}\underline{e}_\theta + \Sigma_{r\varphi}\underline{e}_\varphi)\Big) \qquad [9.10]$$

with

$$\begin{aligned} \Sigma_{rr} &= \Sigma(\cos^2\varphi \sin^2\theta - \cos^2\theta) \\ \Sigma_{r\theta} &= \Sigma \sin\theta\cos\theta(1+\cos^2\varphi) \\ \Sigma_{r\varphi} &= -\Sigma\sin\theta\sin\varphi\cos\varphi \end{aligned} \qquad [9.11]$$

3 Note that r refers to the polar coordinates system on a given crack (and not to spherical coordinates).

We now introduce [9.10] into [9.1]. After integration with respect to r, θ and φ, we obtain:

$$\frac{\mathcal{N}}{4\pi}\int \sin\theta \, d\theta d\varphi \Big(\int_{\mathcal{C}_{\theta,\varphi}} [\![\underline{\xi}]\!] \overset{s}{\otimes} \underline{e}_r(\theta,\varphi) \, dS \Big) = \frac{Q_\mu d}{2\mu^s}\mathbf{\Sigma} \qquad [9.12]$$

with

$$Q_\mu = \frac{32}{45}\frac{(1-\nu^s)(5-\nu^s)}{2-\nu^s} \qquad [9.13]$$

[9.1] eventually reads

$$\mathbf{E} = \left(\frac{1}{2\mu^s} + \frac{Q_\mu d}{2\mu^s}\right)\mathbf{\Sigma} \qquad [9.14]$$

The comparison of [9.14] with [9.2] yields:

$$\mu^{\text{hom}} = \frac{\mu^s}{1+Q_\mu d} \qquad [9.15]$$

Note that this derivation has neglected unilateral effects due to crack closure, which are now examined.

9.1.2. *Stress-based dilute estimate of stiffness with closed cracks*

The present model of closed crack assumes frictionless contact between the two lips of the crack. In other words, there is no tangential stress on the crack lip, as well as no normal displacement jump. Let us again consider the situation of a single crack embedded in an infinite medium (see Chapter 5). When the loading is of mode I type (see section 5.1), it is readily seen that the presence of the crack does not affect the stress and strain states that are uniform throughout the medium. In contrast, when the loading is of the mode II type, the response is identical to that of an opened crack, as derived in section 5.2.

We again consider an isotropic distribution of crack orientations. The homogenized medium is therefore isotropic and its state equation is of the form given in [9.2]. The modulii k^{hom} and μ^{hom} can be determined by the reasoning described in the previous section. It is based on [9.1], which is analyzed successively in the case of a purely isotropic loading and in the case of a pure shear. Let us briefly review the derivation.

In the case of an isotropic loading ($\Sigma 1$), it is readily seen that the presence of closed cracks does not modify the response of the *r.e.v.* and the loading induces no displacement jump. Clearly, [9.1] indicates that the macroscopic response to this very loading is identical to that of the solid. In other words, the effective bulk modulus is equal to that of the solid:

$$k^{\text{kom}} = k^s \qquad [9.16]$$

We now consider a pure shear defined by the macroscopic stress state of [9.9]. Following the previous remarks, it appears that the expression of the displacement jump can be derived from [9.10] in which the normal component (along $\underline{e}_r$) should be removed:

$$[\![\underline{\xi}]\!] = \frac{16(1-\nu^{s2})}{\pi E^s(2-\nu^s)}\sqrt{a^2-r^2}\left(\Sigma_{r\theta}\underline{e}_\theta + \Sigma_{r\varphi}\underline{e}_\varphi\right) \qquad [9.17]$$

with the expressions of $\Sigma_{r\theta}$ and $\Sigma_{r\varphi}$ given in [9.11]. Introducing this expression of $[\![\underline{\xi}]\!]$ into [9.1] yields

$$\frac{\mathcal{N}}{4\pi}\int \sin\theta\, d\theta d\varphi\Big(\int_{\mathcal{C}_{\theta,\varphi}} [\![\underline{\xi}]\!] \overset{s}{\otimes} \underline{e}_r(\theta,\varphi)\, dS\Big) = \frac{Q'_\mu d}{2\mu^s}\boldsymbol{\Sigma} \qquad [9.18]$$

with

$$Q'_\mu = \frac{32}{15}\frac{1-\nu^s}{2-\nu^s} \quad [9.19]$$

and further:

$$\mu^{\text{hom}} = \frac{\mu^s}{1+Q'_\mu d} \quad [9.20]$$

Note (see equation [9.13]) that

$$Q_\mu = \frac{5-\nu^s}{3}Q'_\mu$$

This means that the homogenized shear modulus for closed cracks is greater than for opened cracks.

9.1.3. *Strain-based dilute estimate of stiffness with opened cracks*

The principle of the strain-based dilute approximation is depicted in Figure 9.1 on the right-hand side. Since the medium in which the crack is embedded is homogeneous with stiffness $\mathbb{C}^s$, we observe that the condition $\varepsilon \to \mathbf{E}$ at infinity is equivalent to $\boldsymbol{\sigma} \to \mathbb{C}^s\!:\!\mathbf{E}$ at infinity (the right-hand side of Figure 9.2). It is therefore possible to take advantage of the results discussed in section 9.1.1.

Let us first consider a spherical macroscopic strain state $\mathbf{E} = E\mathbf{1}$. The previous remark indicates that the integral term of [9.1] can be evaluated from [9.6] in which Σ is replaced by $3k^s E$:

$$\frac{\mathcal{N}}{4\pi}\int \sin\theta\, d\theta d\varphi\Big(\int_{\mathcal{C}_{\theta,\varphi}} [\![\underline{\xi}]\!] \overset{s}{\otimes} \underline{e}_r(\theta,\varphi)\, dS\Big) = Q_k d E\mathbf{1} \quad [9.21]$$

In turn, incorporating this result into [9.1] yields:

$$\mathbf{E} = \mathbb{S}^s : \mathbf{\Sigma} + Q_k d\mathbf{E} \qquad [9.22]$$

for any spherical strain state. From the comparison with [9.2], we conclude

$$k^{\text{hom}} = k^s(1 - Q_k d) \qquad [9.23]$$

We now consider a deviatoric strain state

$$\mathbf{E} = E(\underline{e}_x \otimes \underline{e}_x - \underline{e}_z \otimes \underline{e}_z) \qquad [9.24]$$

at infinity. This amounts to prescribe a stress state of the form [9.9] with $\Sigma = 2\mu^s E$ at infinity. Accordingly, it is readily seen from [9.12] that

$$\frac{\mathcal{N}}{4\pi} \int \sin\theta \, d\theta d\varphi \Big(\int_{\mathcal{C}_{\theta,\varphi}} [\![\underline{\xi}]\!] \overset{s}{\otimes} \underline{e}_r(\theta,\varphi) \, dS \Big) = Q_\mu d\mathbf{E} \qquad [9.25]$$

In turn, the introduction of this result into [9.1] gives

$$\mathbf{E} = \mathbb{S}^s : \mathbf{\Sigma} + Q_\mu d\mathbf{E} \qquad [9.26]$$

which holds for any strain state of the form [9.24]. Again, the comparison with [9.2] yields:

$$\mu^{\text{hom}} = \mu^s(1 - Q_\mu d) \qquad [9.27]$$

9.1.4. *Strain-based dilute estimate of stiffness with closed cracks*

Let us briefly repeat the approach of section 9.1.3 in the case of closed cracks. We therefore again focus on a strain-based dilute approximation as depicted in Figure 9.1 on the right-hand side.

We first consider an isotropic macroscopic strain state $\mathbf{E} = E\mathbf{1}$. It appears to be equivalent to an asymptotic stress state $\mathbb{C}^s : \mathbf{E} = 3k^s E\mathbf{1}$. As stated in section 9.1.2, this loading does not induce any displacement jump. Accordingly, [9.1] reads $\mathbf{E} = \mathbb{S}^s : \boldsymbol{\Sigma}$, which in turn yields $k^{\text{hom}} = k^s$.

Consider now a deviatoric macroscopic strain state

$$\mathbf{E} = E(\underline{e}_x \otimes \underline{e}_x - \underline{e}_z \otimes \underline{e}_z) \qquad [9.28]$$

An asymptotic strain state of this form at infinity amounts to the stress state $\boldsymbol{\Sigma} = 2\mu^s \mathbf{E}$ at infinity. Recalling [9.18], we end up with

$$\mathbf{E} = \mathbb{S}^s : \boldsymbol{\Sigma} + Q'_\mu d\mathbf{E} \qquad [9.29]$$

Eventually, the comparison with [9.2] yields:

$$\mu^{\text{hom}} = \mu^s(1 - Q'_\mu d) \qquad [9.30]$$

9.2. A refined strain-based scheme

At the first order in the damage variable d, the stress- and strain-based estimates provided in sections 9.1.1–9.1.4 are consistent with one another. However, the strain-based and stress-based estimates are not strictly identical. In order to understand the origin of the discrepancy, let us move back to Figures 9.1 and 9.2 in which the right-hand side summarizes the idea of the underlying approximation.

In the two problems defined on each side of Figure 9.2, it is readily seen that the average stress[4] in the solid $\overline{\sigma}$ is equal to

4 When the domain is infinite, the average is defined in an asymptotic sense:

$$\overline{\sigma} = \lim_{R\to\infty} \frac{1}{|BR|} \int_{B_R^s} \sigma \, dV$$

where B_R^s is the solid part of the sphere B_R of radius R.

$\mathbf{\Sigma}$ (see section 8.3.3). In this sense, the stress-based dilute scheme is consistent with the stress average rule. This means that the average stress in the solid is identical to the real problem defined on the *r.e.v.* and in the auxiliary problem defined on an infinite space.

Let us now consider the two problems defined in Figure 9.1. For the problem defined on the left side, [8.43] can be applied with Ω being the *r.e.v.* This shows that the macroscopic strain $\mathbf{E}$ and the average strain $\overline{\varepsilon}$ in the solid are *not* identical. In turn, the same equation [8.43] can be applied in an asymptotic sense (Ω being replaced by the sphere of radius R) to the problem defined on the right-hand side of Figure 9.1. Observing that there is a single crack, it is asymptotically obtained (when $R \to \infty$) that $\overline{\varepsilon} = \mathbf{E}$. In this sense, the strain-based dilute scheme is *not* consistent with the strain average rule. In other words, the average strains in the solid in the auxiliary problem, on the one hand, and in the real problem, on the other hand, differ from one another. As such, the strain-based dilute scheme seems less satisfactory than the stress-based dilute scheme.

Let us denote by $\mathbf{E}_o = \overline{\varepsilon}$ the average strain in the solid phase of the *r.e.v.* (real problem). In order to improve the strain-based dilute scheme, the above reasoning suggests that strain-based remote boundary conditions of the auxiliary problem should involve $\mathbf{E}_o$ instead of the macroscopic strain tensor $\mathbf{E}$, as shown in Figure 9.3. $\mathbf{E}_o$ is an unknown of the problem. However, owing to the homogeneity of the solid phase, $\mathbf{E}_o$ is related to the average stress in the solid phase of the *r.e.v.* (real problem):

$$\mathbf{E}_o = \overline{\varepsilon} = \mathbb{S}^s : \overline{\boldsymbol{\sigma}} = \mathbb{S}^s : \mathbf{\Sigma} \qquad [9.31]$$

As in section 9.1.3, we observe that the remote condition $\boldsymbol{\varepsilon} \to \mathbf{E}_o$ at infinity in the auxiliary problem is equivalent to $\boldsymbol{\sigma} \to \mathbb{C}^s : \mathbf{E}_o$ at infinity. Using [9.31], the remote boundary condition now reads $\boldsymbol{\sigma} \to \boldsymbol{\Sigma}$. In other words, the strain-based scheme involving the auxiliairy strain $\mathbf{E}_o$ instead of $\mathbf{E}$ is strictly equivalent to the so-called stress-based dilute scheme.

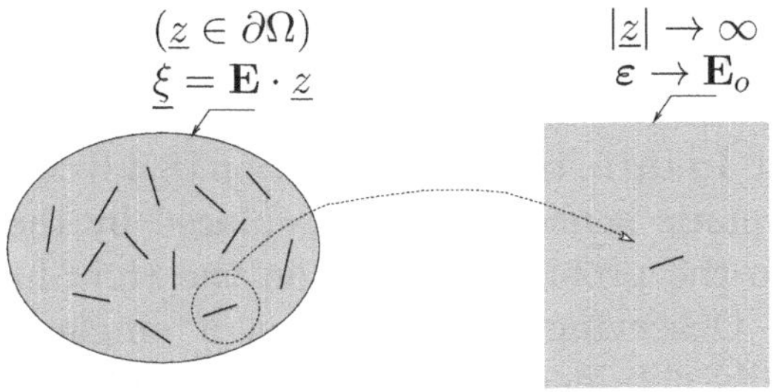

Figure 9.3. *Non-dilute strain-based approximation. The boundary conditions involve the auxiliairy strain tensor* $\mathbf{E}_o = \overline{\varepsilon}$

9.3. Homogenization in plane strain conditions for anisotropic materials

9.3.1. *Opened cracks*

In order to illustrate the homogenization procedure in plane strain conditions, we consider a set of parallel opened cracks embedded in an anisotropic matrix (see also [HOR 83]). We will take the advantage of results established in section 4.2.5. The case of an isotropic matrix will be regarded as a direct application.

Let $\underline{e}_y$ define the unit normal to the cracks of the considered family. The most straightforward way consists of deriving the components of the tensor a^K of [8.54] as well as its average $\overline{\mathrm{a}}^K$ and to apply [8.56] and [8.57].

From [4.81] and [4.83], the displacement jumps are put in the form:

$$\begin{pmatrix} [\![\xi_x]\!] \\ [\![\xi_y]\!] \end{pmatrix} = \begin{pmatrix} a^K_{xx} & a^K_{xy} \\ a^K_{yx} & a^K_{yy} \end{pmatrix} \begin{pmatrix} \sigma^\infty_{xy} \\ \sigma^\infty_{yy} \end{pmatrix} \qquad [9.32]$$

with

$$\begin{aligned} a^K_{xx} &= -2b_{11}\sqrt{a^2 - x^2}\mathcal{R}e(i(\mu_1 + \mu_2)) \\ a^K_{xy} &= -2b_{11}\sqrt{a^2 - x^2}\mathcal{R}e(i\mu_1\mu_2) \\ a^K_{yx} &= +2b_{22}\sqrt{a^2 - x^2}\mathcal{R}e(\frac{i}{\mu_1\mu_2}) \\ a^K_{yy} &= +2b_{22}\sqrt{a^2 - x^2}\mathcal{R}e(i\frac{\mu_1 + \mu_2}{\mu_1\mu_2}) \end{aligned} \qquad [9.33]$$

Let us observe that

$$\mathcal{R}e(i\mu_1\mu_2) = -\mathcal{R}e(i\overline{\mu_1\mu_2}) = -\mathcal{R}e(\frac{i}{\mu_1\mu_2})|\mu_1\mu_2|^2 \qquad [9.34]$$

The symmetry $a^K_{xy} = a^K_{yx}$ is then readily obtained from [1.79].

Let us introduce the crack density $\mathcal{N}$, which denotes the number of cracks par unit surface. Following the definition of $\bar{\mathrm{a}}^K$ of [8.58], it is readily seen from [9.33] that:

$$\begin{aligned} \overline{a}^K_{xx} &= -\frac{\pi a^2}{|S_\Omega|} b_{11}\mathcal{R}e(i(\mu_1 + \mu_2)) \\ \overline{a}^K_{xy} &= -\frac{\pi a^2}{|S_\Omega|} b_{11}\mathcal{R}e(i\mu_1\mu_2) \\ \overline{a}^K_{yy} &= +\frac{\pi a^2}{|S_\Omega|} b_{22}\mathcal{R}e(i\frac{\mu_1 + \mu_2}{\mu_1\mu_2}) \end{aligned} \qquad [9.35]$$

where $|S_\Omega|$ is the cross-section of Ω. The non-vanishing components of $\overline{\mathbb{S}}^K$ as defined in [8.58] are:

$$\overline{S}^K_{1212} = \frac{1}{4}\overline{a}^K_{xx}; \quad \overline{S}^K_{1222} = \frac{1}{2}\overline{a}^K_{xy}; \quad \overline{S}^K_{2222} = \overline{a}^K_{yy} \qquad [9.36]$$

as well as the components derived from above by the usual symmetry rules. Introducing the damage parameter $\delta = \mathcal{N}a^2$, we eventually obtain the expression of the additional compliance due to the considered crack family in the form:

$$\begin{aligned} \sum_K \overline{S}^K_{1212} &= -\frac{\pi\delta}{4} b_{11} \mathcal{R}e(i(\mu_1 + \mu_2)) \\ \sum_K \overline{S}^K_{1222} &= -\frac{\pi\delta}{2} b_{11} \mathcal{R}e(i\mu_1\mu_2) \\ \sum_K \overline{S}^K_{2222} &= +\pi\delta b_{22} \mathcal{R}e(i\frac{\mu_1 + \mu_2}{\mu_1\mu_2}) \end{aligned} \qquad [9.37]$$

These results are to be introduced in [8.57].

For illustrative purposes, let us examine the particular case of an isotropic matrix for which:

$$b_{11} = b_{22} = \frac{1 - \nu^{s2}}{E^s} \qquad [9.38]$$

whereas $\mu_1 = \mu_2 = i$. [9.37] now reads:

$$\begin{aligned} \sum_K \overline{S}^K_{1212} &= \pi\delta \frac{1 - \nu^{s2}}{2E^s} \\ \sum_K \overline{S}^K_{1222} &= 0 \\ \sum_K \overline{S}^K_{2222} &= +2\pi\delta \frac{1 - \nu^{s2}}{E^s} \end{aligned} \qquad [9.39]$$

Introducing the above estimates into [8.57] yields:

$$S^{\text{hom}}_{1212} = \frac{1}{4\mu^s}\left(1 + \pi\delta(1-\nu^s)\right)$$
$$S^{\text{hom}}_{2222} = \frac{1}{E^s}\left(1 + 2\pi\delta(1-\nu^{s2})\right) \qquad [9.40]$$

The other components of the matrix of the tensor are identical to those of $\mathbb{S}^s$. It is very important to recall that the relevancy of the homogenized tensor thus defined is restricted to a stress state $\mathbf{\Sigma}$ such that the associated strain state $\mathbb{S}^{\text{hom}} : \mathbf{\Sigma}$ meets the plane strain requirement. Owing to the structure of $\mathbb{S}^{\text{hom}}$, this amounts to the same condition applied to $\mathbb{S}^s : \mathbf{\Sigma}$. In the isotropic case considered herein, the plane strain conditions expressed on the stress tensor read:

$$\Sigma_{13} = \Sigma_{23} = 0; \quad \Sigma_{33} = \nu^s\left(\Sigma_{11} + \Sigma_{22}\right) \qquad [9.41]$$

9.3.2. *Closed cracks*

We again consider a set of parallel cracks embedded in an anisotropic matrix but now the cracks are assumed to be closed (see section 4.1.5). It is possible to derive the effective compliance in the same way as described in the previous section. Indeed, [9.32] is still valid, except for the fact that the coefficients a^K_{ij} are now given by (see equation [4.94]):

$$a^K_{xx} = -2b_{11}\sqrt{a^2 - x^2} \times \left(\mathcal{R}e(i(\mu_1+\mu_2)) + \mathcal{I}m(\mu_1\mu_2)\frac{\mathcal{R}e(\frac{i}{\mu_1\mu_2})}{\mathcal{R}e(i\frac{\mu_1+\mu_2}{\mu_1\mu_2})}\right) \qquad [9.42]$$

while

$$a^K_{xy} = a^K_{yy} = 0$$

It follows that:

$$\overline{a}_{xx}^{K} = -\frac{\pi a^2}{|S_\Omega|} b_{11} \left(\mathcal{R}e(i(\mu_1 + \mu_2)) + \mathcal{I}m(\mu_1\mu_2) \frac{\mathcal{R}e(\frac{i}{\mu_1\mu_2})}{\mathcal{R}e(i\frac{\mu_1+\mu_2}{\mu_1\mu_2})} \right) \quad [9.43]$$

and the following additional compliance is eventually obtained:

$$\sum_K \overline{S}_{1212}^{K} = -\frac{\pi\delta}{4} b_{11} \times \left(\mathcal{R}e(i(\mu_1 + \mu_2)) + \mathcal{I}m(\mu_1\mu_2) \frac{\mathcal{R}e(\frac{i}{\mu_1\mu_2})}{\mathcal{R}e(i\frac{\mu_1+\mu_2}{\mu_1\mu_2})} \right) \quad [9.44]$$

It is worth noting the fact that S_{1212}^{hom} is only influenced by the presence of closed cracks. Let us finally briefly consider the case of closed cracks in an isotropic matrix. This amounts to let $\mu_1 = \mu_2 = i$ in the above estimate. It is readily seen that S_{1212}^{hom} has the same expression as in the case of opened cracks:

$$S_{1212}^{\text{hom}} = \frac{1}{4\mu^s} \left(1 + \pi\delta(1 - \nu^s)\right) \quad [9.45]$$

all other coefficients being unaffected by the cracks.

10

Eshelby-based Estimates of Strain Concentration and Stiffness

Considering the example of an isotropic behavior, Chapter 9 has shown that homogenized bulk and shear moduli may be derived from the 3D Griffith crack model combined with the solutions derived in Chapter 5. In turn, the present following chapters consider the implementation of the inhomogeneity model for cracks (section 8.2) in a homogenization scheme in detail. The ultimate goal is identical, namely the derivation of estimates for the effective elastic properties (stiffness or compliance). In the present chapter, the uniform strain boundary conditions are adopted (see section 8.2.3). The following chapter will investigate the estimates derived from uniform stress boundary conditions (section 8.2.4).

10.1. Dilute estimate of the strain concentration tensor: general features

10.1.1. *The general case*

The damaging effect related to the presence of the cracks is represented in [8.29] by the term $\varphi_i \overline{\mathbb{A}}^i$. From a practical point of view, we now need to estimate the average strain rate in a crack. The approach presented in this section belongs to the Eshelby-based estimates in the so-called dilute framework. It is based on a geometrical model of cracks in which the latter are described as flat ellipses (plane strain elasticity) or flat spheroids (3D case). The idea of the dilute assumption was already presented in the framework of the Griffith crack model (see section 9.1): it is assumed that the crack density is sufficiently small, so that the cracks do not interact with one another. In other words, this means that any crack in the *r.e.v* reacts as if it were embedded in an infinite medium having the mechanical properties of the solid.

Working now with the inhomogeneity model, the situation corresponds to the classial Eshelby problem. The latter has been presented in the general case in section 2.4 from which the elliptic (section 4.3) and ellipsoidal (Chapter 6) crack models have been derived. The main result is now recalled briefly.

An elliptic (respectively, ellipsoidal) elastic heterogeneity $\mathcal{E}$ with stiffness $\mathbb{C}^{\mathcal{E}}$ is embedded in an infinite homogeneous medium[1] with stiffness $\mathbb{C}^o$. Uniform strain rate boundary conditions are prescribed at infinity, that is:

$$|\underline{z}| \to \infty : \quad \dot{\underline{\xi}}(\underline{z}) \to \dot{\mathbf{E}} \cdot \underline{z} \qquad [10.1]$$

The essential result lies in the fact that the strain rate field induced by this loading in the ellipsoidal heterogeneity is uniform. More precisely[2]:

$$(\forall \underline{z} \in \mathcal{E}) \quad \dot{\varepsilon}(\underline{z}) = \left(\mathbb{I} + \mathbb{P}^{\mathcal{E}}(X) : \delta\mathbb{C}\right)^{-1} : \dot{\mathbf{E}} \qquad [10.2]$$

1 The elastic medium with stiffness $\mathbb{C}^o$ can coincide with the real solid phase. In this case, $\mathbb{C}^o = \mathbb{C}^s$ (see sections 10.3–10.5). However, this may not be the case, as in the self-consistent scheme [ZAO 02, DOR 06].

2 Due to the nonlinearity induced by the geometry of cracks modeled as flat ellipses or ellipsoids, the classical result must be stated in rate form.

$\delta\mathbb{C} = \mathbb{C}^{\mathcal{E}} - \mathbb{C}^o$ represents the contrast of stiffness between the heterogeneity and the medium in which it is embedded. The so-called Hill tensor $\mathbb{P}^{\mathcal{E}}(X)$ only depends on the shape of the heterogeneity $\mathcal{E}$ and on the stiffness tensor $\mathbb{C}^o$. Its coefficients have been given in the elliptic case in sections 2.5.3 (elliptic model) and 2.6.1 (ellipsoidal model).

10.2. The particular case of opened cracks

Let us now implement this result in the case of opened flat cracks in which the stiffness of the heterogeneity is $\mathbb{C}^{\mathcal{E}} = 0$. Introducing the Eshelby tensor $\mathbb{S}^{\mathcal{E}}(X) = \mathbb{P}^{\mathcal{E}}(X) : \mathbb{C}^o$, [10.2] now reads:

$$(\forall \underline{z} \in \mathcal{E}) \quad \dot{\varepsilon}(\underline{z}) = \left(\mathbb{I} - \mathbb{S}^{\mathcal{E}}(X)\right)^{-1} : \dot{\mathbf{E}} \qquad [10.3]$$

$A^{esh}(X) = (\mathbb{I} - \mathbb{S}^{\mathcal{E}})^{-1}$ can be interpreted as the strain rate concentration tensor for the single crack embedded in an infinite medium. The analysis of its coefficients provides a quantitative insight into the large strain mechanism, which modifies the geometry of the crack.

10.2.1. *Spheroidal crack*

To begin with, let us observe that the coefficients A^{esh}_{ijkl} with $i, j \neq 3$ are $\mathcal{O}(1)$. In the case of an isotropic stiffness tensor $\mathbb{C}^o$ (Poisson's ratio ν^o), the leading terms in the series expansions of these coefficients are (see sections 2.6.1 and 2.6.2):

$$\begin{aligned} A^{esh}_{1111} &= 1 - \frac{\nu^o}{2} \\ A^{esh}_{1122} &= -\frac{\nu^o}{2} \\ A^{esh}_{1133} &= \frac{\nu^o - 1}{2} \\ A^{esh}_{1212} &= \frac{1}{2} \end{aligned} \qquad [10.4]$$

This implies that the strain rates in the crack plane, namely $\dot{\varepsilon}_{11}$, $\dot{\varepsilon}_{12}$ and $\dot{\varepsilon}_{22}$ are of the order of $\dot{\mathbf{E}}$. In particular, this means that the variation of the radius a of the crack is negligible.

Let us now examine the series expansions of the remaining coefficients:

$$\begin{aligned} A^{esh}_{3311} &= \frac{1}{X}\frac{4\nu^o(1-\nu^o)}{\pi(1-2\nu^o)} \\ A^{esh}_{3322} &= A^{esh}_{3311} \\ A^{esh}_{3333} &= \frac{1}{X}\frac{4(1-\nu^o)^2}{\pi(1-2\nu^o)} \\ A^{esh}_{2323} &= \frac{1}{X}\frac{2(1-\nu^o)}{\pi(2-\nu^o)} \\ A^{esh}_{1313} &= A^{esh}_{2323} \end{aligned} \qquad [10.5]$$

[10.5] reveals that the local normal strain rate $\dot{\varepsilon}_{33}$ as well as the local shear strain rates $\dot{\varepsilon}_{13}$ and $\dot{\varepsilon}_{23}$ are $\mathcal{O}(1/X)$. In other words, they are by an order of magnitude larger than the strain rate at infinity. This illustrates the strain concentration effect associated with the geometry of flat crack and emphasizes the fact that the strain concentration rule [10.3] is nonlinear in nature and would not be valid if it were written in terms of strain instead of strain rate.

10.2.2. *Elliptic crack*

Similarly, the coefficients A^{esh}_{11kl} are $\mathcal{O}(1)$. The leading terms are (see section 2.5.3):

$$\begin{aligned} A^{esh}_{1111} &= 1-\nu^o \\ A^{esh}_{1122} &= \nu^o - 1 \end{aligned} \qquad [10.6]$$

This implies that the strain rate in the crack plane, namely $\dot{\varepsilon}_{11}$, is of the order of $\dot{\mathbf{E}}$. Again, the variation of the radius a of the crack is negligible. In contrast, the coefficients A^{esh}_{1212}, A^{esh}_{2211} and A^{esh}_{2222} prove to be $\mathcal{O}(1/X)$:

$$A^{esh}_{1212} = \frac{1-\nu^o}{2X}$$

$$A^{esh}_{2211} = \frac{2\nu^o(1-\nu^o)}{(1-2\nu^o)X} \qquad [10.7]$$

$$A^{esh}_{2222} = \frac{2(1-\nu^o)^2}{(1-2\nu^o)X}$$

This means that the normal strain rate $\dot{\varepsilon}_{22}$ as well as the shear strain rate $\dot{\varepsilon}_{12}$ are by an order of magnitude larger than the strain rate $\dot{\mathbf{E}}$ at infinity.

In the framework of plane strains, $\dot{\mathbf{E}}$ has no component $\dot{E}_{i3}$ and, in turn, the local strain concentration tensor has no component A_{i3kl} and A_{ij3l}. In fact, it is interesting to extend the result stated in [10.7] in considering the 3D problem of a cylindrical crack along the axis parallel to $\underline{e}_3$, with elliptic cross-section in the plane $(\underline{e}_1, \underline{e}_2)$. The 3D response to a loading of the form:

$$\dot{\mathbf{E}} = \dot{E}_{11}\underline{e}_1 \otimes \underline{e}_1 + \dot{E}_{22}\underline{e}_2 \otimes \underline{e}_2 + \dot{E}_{12}\left(\underline{e}_1 \otimes \underline{e}_2 + \underline{e}_2 \otimes \underline{e}_1\right)$$

would be obviously the one described by [10.6] and [10.7]. In contrast, a loading of the form

$$\dot{\mathbf{E}} = \dot{E}_{33}\underline{e}_3 \otimes \underline{e}_3 + \dot{E}_{23}(\underline{e}_2 \otimes \underline{e}_3 + \underline{e}_3 \otimes \underline{e}_2) + \dot{E}_{13}\left(\underline{e}_1 \otimes \underline{e}_3 + \underline{e}_3 \otimes \underline{e}_1\right)$$

would induce additional local strain (rate) components $\dot{\varepsilon}_{13}$, $\dot{\varepsilon}_{23}$ and $\dot{\varepsilon}_{33}$. It can be shown that only A^{esh}_{2233} and A^{esh}_{2323} are of the order of $\mathcal{O}(1/X)$:

$$A^{esh}_{2233} = \frac{1}{X}\frac{2\nu^o(1-\nu^o)}{1-2\nu^o}; \quad A^{esh}_{2323} = \frac{1}{2X} \qquad [10.8]$$

10.2.3. *Crack opening change*

The fact that there is a strain concentration effect in the direction normal to the crack implies that the opening of the crack is subjected to large variations. This holds for both the spheroidal and elliptic models. The derivation of the crack opening variation as a function of the loading is an instructive application of [10.3]. The starting point consists of observing that the normal strain rate $\dot{\varepsilon}_{nn}$ is equal to the ratio $\dot{c}/c$. This yields:

$$\frac{\dot{c}}{a} = \underline{n} \cdot \left(X(\mathbb{I} - \mathbb{S})^{-1} : \dot{\mathbf{E}} \right) \cdot \underline{n} \qquad [10.9]$$

Since $X \ll 1$, the product $X(\mathbb{I} - \mathbb{S})^{-1}$ can be replaced by its limit $\mathbb{T}$ when $X \to 0$:

$$X(\mathbb{I} - \mathbb{S}(X))^{-1} \approx \mathbb{T} = \lim_{X \to 0} X(\mathbb{I} - \mathbb{S}(X))^{-1} \qquad [10.10]$$

Observing that $\mathbb{T}$ is a constant, the integration of [10.9] with respect to time yields:

$$\frac{c - c_o}{a} = \underline{n} \cdot \left(\mathbb{T} : \mathbf{E} \right) \cdot \underline{n} \qquad [10.11]$$

where c and c_o are, respectively, the current and initial values of the half-opening. The variation $\delta c = c - c_o$ of the half-opening thus appears as a linear function of the strain $\mathbf{E}$ at infinity. Crack closure being characterized by $\delta c = -c_o$, [10.11] allows to characterize the threshold beyond which the crack lips are in contact.

The tensor $\mathbb{T}$ defined by [10.10] plays an important role in the micromechanics approach to the elastic properties of microcracked media. It only depends on the stiffness tensor $\mathbb{C}^o$ and on the penny-shaped crack orientation. In the case of an isotropic solid, its non-vanishing coefficients are readily derived from [10.5] in the framework of the spheroidal model:

$$\begin{aligned} T_{3311} &= \frac{4\nu^o(1-\nu^o)}{\pi(1-2\nu^o)} \\ T_{3322} &= T_{3311} \\ T_{3333} &= \frac{4(1-\nu^o)^2}{\pi(1-2\nu^o)} \\ T_{2323} &= \frac{2(1-\nu^o)}{\pi(2-\nu^o)} \\ T_{1313} &= T_{2323} \end{aligned} \qquad [10.12]$$

with the normal to the crack parallel to $\underline{e}_3$. Similarly, in the framework of the elliptic model, the non-vanishing coefficients of $\mathbb{T}$ are:

$$\begin{aligned} T_{2211} &= \frac{2\nu^o(1-\nu^o)}{1-2\nu^o} \\ T_{2222} &= \frac{2(1-\nu^o)^2}{1-2\nu^o} \\ T_{1212} &= \frac{1-\nu^o}{2} \end{aligned} \qquad [10.13]$$

with the normal to the crack parallel to $\underline{e}_2$.

It is recalled that [10.13] is dedicated to plane strain conditions. The extension of this result to the complete 3D response in the presence of a cylindrical crack[3] requires two additional components, namely T_{2233} and T_{2323}, readily derived from [10.8]:

$$T_{2233} = \frac{2\nu^o(1-\nu^o)}{1-2\nu^o}; \quad T_{2323} = \frac{1}{2} \qquad [10.14]$$

3 Axis along $\underline{e}_3$, cross-section in the plane $(\underline{e}_1, \underline{e}_2)$, normal $\underline{e}_2$.

In view of forthcoming use, let us introduce the "plane strain" tensor $\mathbb{T}^{ps}$ defined by [10.13] and the 3D tensor $\mathbb{T}^{cyl}$ for the same cylindrical geometry of the crack accounting for any loading:

$$\mathbb{T}^{cyl} = \mathbb{T}^{ps} + T_{2233}\underline{e}_2 \otimes \underline{e}_2 \otimes \underline{e}_3 \otimes \underline{e}_3 + 4T_{2323}(\underline{e}_2 \overset{s}{\otimes} \underline{e}_3) \otimes (\underline{e}_3 \overset{s}{\otimes} \underline{e}_2) \quad [10.15]$$

10.3. Dilute estimates of the effective stiffness for opened cracks

It was shown by [8.29] that an estimate of the effective stiffness tensor $\mathbb{C}^{\text{hom}}$ is necessarily based on an estimate of the strain (rate) concentration tensor in the cracks. We begin with the so-called dilute homogenization scheme, which neglects the mechanical interaction between cracks. An attempt to capture the latter is presented in section 10.5. If the mechanical interaction between cracks in the *r.e.v* can be neglected, we have seen in section 10.1 that each crack reacts to the macroscopic loading as if it were embedded in an homogenous medium having the same stiffness as the solid itself. Accordingly, the strain rate concentration rule [8.30] can be approximated by the solution [10.3] of the Eshelby problem. Hence, considering crack n°i, the average $\overline{\mathbb{A}}^i$ in [8.29] can be estimated by $\left(\mathbb{I} - \mathbb{S}_i(X_i)\right)^{-1}$, depending on its orientation and aspect ratio X_i. Hence, the elementary contribution of crack n°i of volume fraction φ_i to the effective stiffness is $-\varphi_i\mathbb{C}^s : \left(\mathbb{I} - \mathbb{S}_i\right)^{-1}$.

In the 3D case, the volume fraction of an oblate spheroid with radius a_i is $\varphi_i = 4\pi a_i^3 X_i/(3|\Omega|)$. In turn, in the 2D case, it takes the form $\varphi_i = \pi a_i^2 X_i/|\mathcal{S}_\Omega|$, where $|\mathcal{S}_\Omega|$ is the cross-section of the *rev*. With these notations, the dilute approximation yields the following estimate of the tangent stiffness tensor $\mathbb{C}^{\text{hom}}$:

$$
\begin{aligned}
3D : \mathbb{C}_{\text{dil}}^{\text{hom},3D} &= \mathbb{C}^s : \left(\mathbb{I} - \frac{4\pi}{3|\Omega|} a_i^3 \mathbb{T}_i\right) \\
2D : \mathbb{C}_{\text{dil}}^{\text{hom},2D} &= \mathbb{C}^s : \left(\mathbb{I} - \frac{\pi}{|S_\Omega|} a_i^2 \mathbb{T}_i\right)
\end{aligned} \qquad [10.16]
$$

Whatever the distribution of crack orientations, [10.16] reveals that the dilute estimate of the tangent stiffness tensor is in fact a constant, that is it is not affected by the geometry changes in the cracks. In other words, despite the geometrical nonlinearity of the strain concentration at the microscopic scale, namely that of each individual crack, the macroscopic state equation appears to be linear. Indeed, [10.2] can be integrated in the form $\boldsymbol{\Sigma} = \mathbb{C}_{\text{dil}}^{\text{hom}} : \mathbf{E}$. This conclusion holds as long as all cracks remain opened.

As far as the use of [10.16] in 2D (plane strain) applications is concerned, the question arises as to which tensor $\mathbb{T}$ should be considered among $\mathbb{T}^{cyl}$ and $\mathbb{T}^{ps}$ (see equation [10.15]). At first sight, the straightforward approach is to use the 3D cylindrical model of crack, that is to select $\mathbb{T} = \mathbb{T}^{cyl}$ and to restrict the use of the corresponding homogenized stiffness tensor to plane strain tensors ($E_{i3} = 0$). Nevertheless, since $\mathbb{C}_{\text{dil}}^{\text{hom},2D}$ operates on macroscopic strain (or strain rate) tensors meeting the plane strain requirements $E_{i3} = 0$, the use of $\mathbb{T}^{cyl}$ rather than $\mathbb{T}^{ps}$ makes no difference in the present case. This means that the 2D determination of $\mathbb{T}^{ps}$ that goes back to that of the two-dimensional Hill tensor as mentioned in section 2.5, is sufficient. This observation is clearly valuable from a pratical point of view in so far as the complete (i.e. 3D) determination of $\mathbb{T}$ can be spared.

Notwithstanding the fact that both choices yield identical results, the use of $\mathbb{T}^{ps}$ deserves a comment with regard to the issue of symmetry. In fact, it is readily seen that the symmetry of the matrix of the tensor $\mathbb{C}_{\text{dil}}^{\text{hom},2D}$ is ensured if $\mathbb{T} = \mathbb{T}^{cyl}$ is selected, whereas symmetry is lost if $\mathbb{T} = \mathbb{T}^{ps}$. In the later case, the tensor is indeed symmetric on the vector subspace of plane strain tensors.

In order to apply the general expression [10.16], an assumption concerning the distribution of crack orientations is required.

10.3.1. *Opened parallel cracks*

In the case of parallel cracks, the tensor $\mathbb{T}$ is identical for all cracks, so that [10.16] reads:

$$\begin{aligned} 3D : \mathbb{C}_{\text{dil}}^{\text{hom}} &= \mathbb{C}^s : (\mathbb{I} - \frac{4\pi}{3|\Omega|}\mathbb{T}\sum_i a_i^3) \\ 2D : \mathbb{C}_{\text{dil}}^{\text{hom}} &= \mathbb{C}^s : (\mathbb{I} - \frac{\pi}{|S_\Omega|}\mathbb{T}\sum_i a_i^2) \end{aligned} \qquad [10.17]$$

10.3.1.1. *3D case*

It is convenient to introduce the crack density $\mathcal{N}$ defined as the number of cracks per unit of volume and the average cube of the crack radius:

$$\overline{a^3} = \frac{1}{\mathcal{N}|\Omega|}\sum_i a_i^3 \qquad [10.18]$$

With these notations, [10.17] now reads:

$$3D : \quad \mathbb{C}_{\text{dil}}^{\text{hom}} = \mathbb{C}^s : (\mathbb{I} - \frac{4\pi d}{3}\mathbb{T}) \qquad [10.19]$$

where $d = \mathcal{N}\overline{a^3}$ is the crack density parameter initially introduced in [BUD 76]. The expressions of the components of tensor $\mathbb{T}$ have been given in [10.12]. It is recalled that $\mathbb{T}$ does not have the symmetry properties of a tensor of elasticity.

Nevertheless, it can be readily verified that $\mathbb{C}_{\text{dil}}^{\text{hom}}$ defined by [10.19] is indeed symmetric.

From [10.19], it is clear that d can also be interpreted as internal damage variable, the evolution of which will be considered in section 13.1.

Clearly enough, the presence of the set of parallel cracks modifies the material symmetries w.r.t those of the solid. If the solid is isotropic, the cracked medium appears as a transverse isotropic material, the direction of symmetry of the anisotropy being the normal to the cracks.

10.3.1.2. *2D case*

In the 2D case, we similarly introduce the crack density $\mathcal{N}$ defined as the number of cracks per unit of surface and the average square of the crack radius:

$$\overline{a^2} = \frac{1}{\mathcal{N} S_\Omega} \sum_i a_i^2 \qquad [10.20]$$

[10.17] now reads:

$$2\text{D}: \quad \mathbb{C}_{\text{dil}}^{\text{hom}} = \mathbb{C}^s : \left(\mathbb{I} - \pi \delta\, \mathbb{T}\right) \qquad [10.21]$$

where $\delta = \mathcal{N}\overline{a^2}$ is the 2D crack density parameter or damage variable. The expressions of the coefficients of tensor $\mathbb{T} = \mathbb{T}^{ps}$ have been given in [10.13].

As opposed to the stiffness tensor defined by [10.19], the matrix of the stiffness tensor of [10.21] (with $\mathbb{T} = \mathbb{T}^{ps}$) is not symmetric. However, when we restrict the use of the tensor of [10.21] to the vector subspace of strain tensors meeting the conditions of plane strain, the symmetry requirement is indeed achieved:

$$(\forall \mathbf{E}, \mathbf{E}')\ E_{i3} = E'_{i3} = 0 \Rightarrow \mathbf{E} : \mathbb{C}_{\text{dil}}^{\text{hom}} : \mathbf{E}' = \mathbf{E}' : \mathbb{C}_{\text{dil}}^{\text{hom}} : \mathbf{E} \qquad [10.22]$$

10.3.2. *Opened randomly oriented cracks*

A simple analytical expression can also been derived when the distribution of crack orientations is isotropic.

10.3.2.1. *3D case*

Let us first consider the 3D case. When all cracks have the same radius a, we introduce the damage parameter $d = \mathcal{N}a^3$; [10.16] then takes the form:

$$\mathbb{C}_{\text{dil}}^{\text{hom}} = \mathbb{C}^s : \left(\mathbb{I} - \frac{4\pi d}{3} < \mathbb{T} > \right) \quad \text{with}$$

$$< \mathbb{T} > = \int_0^{\pi} d\theta \int_0^{2\pi} \mathbb{T}(\theta, \psi) \frac{\sin\theta}{4\pi} d\psi \qquad [10.23]$$

where $\mathbb{T}(\theta, \psi)$ refers to the crack family, the normal of which is defined on the unit sphere by the angles θ and ψ. The average $< \mathbb{T} >$ of $\mathbb{T}(\theta, \psi)$ over the unit sphere is an isotropic tensor. Introducing the fourth-order tensors $\mathbb{J}$ and $\mathbb{K}$ defined as $J_{ijkl} = \delta_{ij}\delta_{kl}/3$ and $\mathbb{K} = \mathbb{I} - \mathbb{J}$, it is found that

$$\mathbb{Q} = \frac{4\pi}{3} < \mathbb{T} >= Q_k \mathbb{J} + Q_\mu \mathbb{K} \qquad [10.24]$$

with[4]

$$Q_k = \frac{16}{9} \frac{1 - \nu^{s2}}{1 - 2\nu^s}; \quad Q_\mu = \frac{32}{45} \frac{(1 - \nu^s)(5 - \nu^s)}{2 - \nu^s} \qquad [10.25]$$

It follows that microcracks-induced damage affects the bulk and shear moduli in a different manner:

$$k_{\text{dil}}^{\text{hom}} = k^s(1 - Q_k d); \quad \mu_{\text{dil}}^{\text{hom}} = \mu^s(1 - Q_\mu d) \qquad [10.26]$$

4 Same definitions as in [9.8] and [9.13].

These results are identical to the strain-based dilute estimates derived with the Griffith crack model (see equations [9.23] and [9.27]). The homogenized Young's modulus and Poisson ratio derived from [10.26] read:

$$\begin{aligned} E_{\text{dil}}^{\text{hom}} &= E^s\left(1 - Q_E d\right) \\ \nu_{\text{dil}}^{\text{hom}} &= \nu^s\left(1 - Q_\nu d\right) \end{aligned} \quad \text{with} \quad \begin{aligned} Q_E &= \frac{16(1-\nu^{s2})(10-3\nu^s)}{45(2-\nu^s)} \\ Q_\nu &= \frac{16(1-\nu^{s2})(3-\nu^s)}{15(2-\nu^s)} \end{aligned} \qquad [10.27]$$

Note that the Poisson ratio is found to be a decreasing function of the damage parameter. The affine expressions of $E_{\text{dil}}^{\text{hom}}$ and $\nu_{\text{dil}}^{\text{hom}}$ as functions of d are first-order expansions of these quantities, according to the dilute approximation.

10.3.2.2. *2D case*

We introduce an orthormal frame $(\underline{e}_1, \underline{e}_2, \underline{e}_3)$ and we consider a set of cracks that are infinite in the direction of $\underline{e}_3$. The behavior of the microcracked material is sought in plane strain elasticity parallel to $(\underline{e}_1, \underline{e}_2)$. Again, under the assumption that all cracks have the same radius a, we introduce $\delta = \mathcal{N} a^2$ so that [10.16] becomes

$$\mathbb{C}_{\text{dil}}^{\text{hom}} = \mathbb{C}^s : \left(\mathbb{I} - \pi\delta < \mathbb{T} > \right) \text{ with } < \mathbb{T} >= \int_0^{2\pi} \mathbb{T}(\theta)\, \frac{d\theta}{2\pi} \qquad [10.28]$$

where $\mathbb{T}(\theta)$ refers to the crack family the normal of which makes the angle θ with $\underline{e}_2$. The relevant components[5] of the average $< \mathbb{T} >$ of $\mathbb{T}(\theta)$ over $\theta \in [0, 2\pi]$ are

$$\begin{aligned} < T >_{1111} =< T >_{2222} &= \frac{(1-\nu^s)^2}{1-2\nu^s} \\ < T >_{2211} =< T >_{1122} &= \frac{(1-\nu^s)\nu^s}{1-2\nu^s} \\ < T >_{1212} &= \frac{1-\nu^s}{2} \end{aligned} \qquad [10.29]$$

5 The coefficients $< T >_{ijkl}$ with $k = 3$ or $l = 3$ are not relevant in the plane strain framework.

The homogenized microcracked material is transversely isotropic w.r.t. the direction $\underline{e}_3$. Only the Young's modulus and the Poisson coefficient associated with the plane $(\underline{e}_1, \underline{e}_2)$ are affected by the cracks. Up to the first order in δ, it is found that:

$$\begin{array}{l} E_{\rm dil}^{\rm hom} = E^s(1 - Q\delta) \\ \nu_{\rm dil}^{\rm hom} = \nu^s(1 - Q\delta) \end{array} \quad \text{with} \quad Q = (1 - \nu^{s2})\pi \qquad [10.30]$$

10.4. Dilute estimates of the effective stiffness for closed cracks

We now consider the case of a closed crack. We assume that the two lips of the crack are in frictionless contact. This means that the crack transmits the normal compressive forces, whereas the shear stress in the crack plane remains equal to 0. Following [DEU 02], it will prove very useful to model such a closed crack by a fictitious flat ellipsoidal inhomogeneity, filled by a fictitious linear elastic isotropic material with a bulk modulus $k^c \neq 0$ and a shear modulus $\mu^c = 0$:

$$\mathbb{C}^c = 3k^c \mathbb{J} \qquad [10.31]$$

As far as the determination of the effective stiffness is concerned, it can be shown that the value of k^c is unimportant, provided that it is not equal to 0. In the sequel, we let k^c be equal to the bulk modulus k^s of the solid, in which stiffness tensor is assumed to be isotropic. Similarly, the aspect ratio X of this fictitious ellipsoid is irrelevant, provided that $X \ll 1$. The strain rate concentration rule [10.2] now reads:

$$\dot{\varepsilon}(\underline{z}) = \left(\mathbb{I} - \mathbb{S} : \mathbb{K}\right)^{-1} : \dot{\mathbf{E}} \qquad [10.32]$$

which replaces [10.3]. Following the same reasoning as for opened cracks, it appears that the elementary contribution of crack n°i with radius a_i and aspect ratio X_i to the effective stiffness [8.27] is $-\varphi_i \mathbb{C}^s : \mathbb{K} : \left(\mathbb{I} - \mathbb{S}_i : \mathbb{K}\right)^{-1}$, so that [10.16] can be replaced by

$$\begin{aligned} 3D &: \mathbb{C}_{\text{dil}}^{\text{hom}} = \mathbb{C}^s : (\mathbb{I} - \frac{4\pi}{3|\Omega|} a_i^3 \mathbb{T}_i') \\ 2D &: \mathbb{C}_{\text{dil}}^{\text{hom}} = \mathbb{C}^s : (\mathbb{I} - \frac{\pi}{|S_\Omega|} a_i^2 \mathbb{T}_i') \end{aligned} \qquad [10.33]$$

where the $\mathbb{T}_i'$ are defined as:

$$\mathbb{T}' = \lim_{X \to 0} X \, \mathbb{K} : \left(\mathbb{I} - \mathbb{S}(X) : \mathbb{K}\right)^{-1} \qquad [10.34]$$

Let us consider the case of an isotropic solid. In the 3D case, the only coefficients of $\mathbb{T}'$ that are not equal to 0 are, as expected, associated with shear strain involving a direction of the crack plane and the normal ($\underline{n} = \underline{e}_3$) to the later:

$$3D: \quad T'_{1313} = T'_{2323} = \frac{2(1-\nu^s)}{\pi(2-\nu^s)} \qquad [10.35]$$

Similarly, in the 2D case, the non-vanishing coefficient of $\mathbb{T}'$ are associated with shear strain involving a direction of the crack line ($\underline{t} = \underline{e}_1$) and the normal ($\underline{n} = \underline{e}_2$) to the later:

$$2D: \quad T'_{1212} = \frac{1-\nu^s}{2} \qquad [10.36]$$

Once the cracks are closed, [10.33] shows that the effective tangent stiffness tensor is constant. Nevertheless, the transition from opened cracks to closed cracks is associated with a stiffening of the *r.e.v*.

10.4.1. *Closed parallel cracks*

Clearly, if the cracks are parallel and have the same radius a, [10.33] becomes:

$$\begin{aligned} 3\text{D} : \mathbb{C}^{\text{hom}}_{\text{dil}} &= \mathbb{C}^s : \left(\mathbb{I} - \frac{4\pi d}{3}\mathbb{T}'\right) \\ 2\text{D} : \mathbb{C}^{\text{hom}}_{\text{dil}} &= \mathbb{C}^s : \left(\mathbb{I} - \pi\delta\mathbb{T}'\right) \end{aligned} \qquad [10.37]$$

10.4.2. *Closed randomly oriented cracks*

In the case of an isotropic distribution of crack orientations, it suffices to replace $< \mathbb{T} >$ by $< \mathbb{T}' >$ in [10.23] (respectively, [10.28]). In the 3D case, if the solid is isotropic, the damaged material is isotropic as well. It is found that

$$< \mathbb{T}' > = \frac{8(1-\nu^s)}{5\pi(2-\nu^s)}\mathbb{K} \qquad [10.38]$$

Hence, as opposed to the case of opened cracks, the bulk modulus is not affected by the damage. Furthermore, the effect on the shear modulus is different from the effect produced by randomly oriented *opened* cracks:

$$k^{\text{hom}}_{\text{dil}} = k^s; \quad \mu^{\text{hom}}_{\text{dil}} = \mu^s\left(1 - Q'_\mu d\right) \qquad [10.39]$$

with

$$Q'_\mu = \frac{32}{15}\frac{1-\nu^s}{2-\nu^s} \qquad [10.40]$$

The expression of the homogenized Young's modulus and Poisson ratio, which can be derived from [10.39], reads:

$$\begin{aligned} E^{\text{hom}}_{\text{dil}} &= E^s\left(1 - Q'_E d\right) \\ \nu^{\text{hom}}_{\text{dil}} &= \nu^s + \frac{32}{45}\frac{(1-\nu^{s2})(1-2\nu^s)}{2-\nu^s}d \end{aligned} \qquad [10.41]$$

with

$$Q'_E = \frac{64}{45}\frac{(1-\nu^{s2})}{2-\nu^s} = \frac{2(1+\nu^s)}{3}Q'_\mu \qquad [10.42]$$

As opposed to what had been obtained for opened cracks (see equation [10.27]), [10.41] indicates that $\nu^{\text{hom}}_{\text{dil}}$ is an increasing function of d.

In the 2D case, assuming again that the solid is isotropic, the damaged material is isotropic with respect to the direction $\underline{e}_3$ in which the cracks are infinite. $< \mathbb{T}' >$ is obtained from [10.36] in the line of [10.28]:

$$2D: \quad \begin{aligned} &< T' >_{1111} = < T' >_{2222} = \frac{1-\nu^s}{4} \\ &< T' >_{1122} = < T' >_{2211} = -\frac{1-\nu^s}{4} \\ &< T' >_{1212} = \frac{1-\nu^s}{4} \end{aligned} \qquad [10.43]$$

As for opened cracks in [10.30], it is possible to derive an expression of the Young's modulus and Poisson ratio characterizing the plane $(\underline{e}_1, \underline{e}_2)$, to the first order with respect to δ. From [10.43], it follows that:

$$\begin{aligned} E^{\text{hom}}_{\text{dil}} &= E^s\big(1 - \frac{Q}{4}\delta\big) \\ \nu^{\text{hom}}_{\text{dil}} &= \nu^s + \frac{Q}{4}(1-\nu^s)\delta \end{aligned} \qquad [10.44]$$

with Q given in [10.30]. Similarly to the 3D case, note that $\nu^{\text{hom}}_{\text{dil}}$ is an increasing function of the damage parameter δ.

10.5. Mori–Tanaka estimate of the effective stiffness

Since it neglects the mechanical interaction between cracks, the dilute scheme is, by nature, restricted to small values of the crack density. The so-called Mori–Tanaka scheme [MOR 73, BEN 87] is an attempt to overcome these shortcomings.

As the dilute scheme, the Mori–Tanaka scheme is based on an Eshelby problem (section 10.1). However, the boundary conditions at infinity are modified. Instead of being defined from the macroscopic strain rate $\dot{\mathbf{E}}$, an auxiliary strain rate $\dot{\mathbf{E}}_o$ is introduced:

$$|\underline{z}| \to \infty : \quad \dot{\underline{\xi}}(\underline{z}) \to \dot{\mathbf{E}}_o \cdot \underline{z} \qquad [10.45]$$

The average strain rate in crack $\mathcal{C}_i$ thus reads:

$$\overline{\dot{\varepsilon}}^i = \left(\mathbb{I} + \mathbb{P}^i : (\mathbb{C}^i - \mathbb{C})\right)^{-1} : \dot{\mathbf{E}}_o \qquad [10.46]$$

As in the dilute scheme, $\mathbb{P}^i$ is the Hill's tensor of a spheroid with the same orientation and aspect ratio as $\mathcal{C}_i$ while $\mathbb{C}^i$ is equal to 0 or to $3k^s\mathbb{J}$, depending whether the crack is opened or closed.

Moreover, it is assumed that the average strain rate in the solid phase of the *r.e.v* is equal to that in the solid phase surrounding the crack in the Eshelby problem with boundary conditions [10.45], which is clearly equal to $\dot{\mathbf{E}}_o$.

$$\overline{\dot{\varepsilon}}^s = \dot{\mathbf{E}}_o \qquad [10.47]$$

The strain average rule $\overline{\dot{\varepsilon}} = \dot{\mathbf{E}}$ now relates $\dot{\mathbf{E}}_o$ to the real macroscopic strain rate $\dot{\mathbf{E}}$:

$$\dot{\mathbf{E}}_o = \left(\varphi_i\left(\mathbb{I} + \mathbb{P}^i : (\mathbb{C}^i - \mathbb{C}^s)\right)^{-1} + (1-\varphi)\mathbb{I}\right)^{-1} : \dot{\mathbf{E}} \qquad [10.48]$$

Introducing [10.48] into [10.46] yields a new estimate of the average strain concentration tensor[6]:

$$\overline{\mathbb{A}}^{i} = \left(\mathbb{I} + \mathbb{P}^{i}:(\mathbb{C}^{i} - \mathbb{C})\right)^{-1} : \left(\varphi_j\left(\mathbb{I} + \mathbb{P}^{j} : (\mathbb{C}^{j} - \mathbb{C})\right)^{-1} + (1-\varphi)\mathbb{I}\right)^{-1} \tag{10.49}$$

After some algebra, the Mori–Tanaka estimate of the tangent homogenized stiffness tensor is then derived from [8.29]:

$$\mathbb{C}^{\text{hom}}_{\text{mt}} = (1-\varphi)\mathbb{C}^{s} : \left(\varphi_i\left(\mathbb{I} + \mathbb{P}^{i} : (\mathbb{C}^{i} - \mathbb{C}^{s})\right)^{-1} + (1-\varphi)\mathbb{I}\right)^{-1} \tag{10.50}$$

In the general case, it is recalled that [10.50] includes a sum on the index i, that is on the whole crack system.

10.5.1. *Opened cracks*

For opened cracks, we have $\mathbb{C}^{i} = 0$. Furthermore, we recall that the elementary crack volume fraction φ_i can be put in the form $\varphi'_i X_i$, where φ'_i is equal to $4\pi a_i^3/(3|\Omega|)$ in the 3D case and to $\pi a_i^2/|\mathcal{S}_\Omega|$ in the 2D case; [10.50] now takes the form:

$$\mathbb{C}^{\text{hom}}_{\text{mt}} = (1-\varphi)\mathbb{C}^{s} : \left(\varphi'_i \mathbb{T}_i + (1-\varphi)\mathbb{I}\right)^{-1} \tag{10.51}$$

Finally, the total crack porosity φ being infinitesimal, [10.51] can be replaced by its limit for $\varphi \to 0$. Accordingly, the Mori–Tanaka counterpart of [10.16] is found to be:

$$\begin{aligned} 3\text{D} : \mathbb{C}^{\text{hom}}_{\text{mt}} &= \mathbb{C}^{s} : \left(\mathbb{I} + \frac{4\pi}{3|\Omega|} a_i^3 \mathbb{T}_i\right)^{-1} \\ 2\text{D} : \mathbb{C}^{\text{hom}}_{\text{mt}} &= \mathbb{C}^{s} : \left(\mathbb{I} + \frac{\pi}{|\mathcal{S}_\Omega|} a_i^2 \mathbb{T}_i\right)^{-1} \end{aligned} \tag{10.52}$$

6 This expression of $\overline{\mathbb{A}}^{i}$ involves a sum on the index j.

We now assume that the crack radii can be approximated by the constant a. In the case of parallel cracks, the above expression yields:

$$\begin{aligned} 3\mathrm{D}: \mathbb{C}_{\mathrm{mt}}^{\mathrm{hom}} &= \mathbb{C}^s : \left(\mathbb{I} + \frac{4\pi d}{3}\mathbb{T}\right)^{-1} \\ 2\mathrm{D}: \mathbb{C}_{\mathrm{mt}}^{\mathrm{hom}} &= \mathbb{C}^s : \left(\mathbb{I} + \pi\delta\mathbb{T}\right)^{-1} \end{aligned} \qquad [10.53]$$

which appears as the generalization of [10.19] and [10.21] to be used when the condition $d \ll 1$ (respectively, $\delta \ll 1$) is not satisfied. With regard to the 2D estimate $\mathbb{C}_{\mathrm{mt}}^{\mathrm{hom},2D} = \mathbb{C}^s : \left(\mathbb{I} + \pi\delta\mathbb{T}\right)^{-1}$, it is not obvious at first sight that both choices $\mathbb{T} = \mathbb{T}^{cyl}$ and $\mathbb{T} = \mathbb{T}^{ps}$ are equivalent, which is indeed the case if the restriction of $\mathbb{C}_{\mathrm{mt}}^{\mathrm{hom,2D}}$ to the vector subspace of plane strain tensors is considered. As already mentioned in section 10.3, the lack of symmetry of $\mathbb{C}_{\mathrm{mt}}^{\mathrm{hom,2D}}$ observed with $\mathbb{T} = \mathbb{T}^{ps}$ is only apparent: in this case, $\mathbb{C}_{\mathrm{mt}}^{\mathrm{hom,2D}}$ is meaningless if the plane strain requirement E_{i3} is not fulfilled.

Similarly, for an isotropic distribution of crack orientations, the dilute estimates [10.23] in 3D (respectively, [10.28] in 2D) should be replaced by:

$$\begin{aligned} 3D: \mathbb{C}_{\mathrm{mt}}^{\mathrm{hom}} &= \mathbb{C}^s : \left(\mathbb{I} + \frac{4\pi d}{3} < \mathbb{T} > \right)^{-1} \\ 2D: \mathbb{C}_{\mathrm{mt}}^{\mathrm{hom}} &= \mathbb{C}^s : \left(\mathbb{I} + \pi\delta < \mathbb{T} > \right)^{-1} \end{aligned} \qquad [10.54]$$

In the 3D case, the Mori–Tanaka estimates of the homogenized bulk and shear moduli read:

$$k_{\mathrm{mt}}^{\mathrm{hom}} = \frac{k^s}{1 + Q_k d}; \quad \mu_{\mathrm{mt}}^{\mathrm{hom}} = \frac{\mu^s}{1 + Q_\mu d} \qquad [10.55]$$

with Q_k and Q_μ given in [10.25]. These expressions should be compared to the homologous quantities derived in the dilute case (see equation [10.26]). They are identical to both the stress-based dilute estimates and the refined strain-based dilute estimates derived with the Griffith crack model (see equations [9.8] and [9.15]). The corresponding Young modulus and Poisson ratio take the form:

$$E_{\mathrm{mt}}^{\mathrm{hom}} = \frac{E^s}{1+Q_E d} \qquad \nu_{\mathrm{mt}}^{\mathrm{hom}} = (1+\nu^s)\frac{1+Q_\mu d}{1+Q_E d} - 1 \tag{10.56}$$

with Q_E given in [10.27].

In the 2D case, the Young's modulus and Poisson ratio in the plane perpendicular to the direction $\underline{e}_3$ in which the cracks are infinite read:

$$E_{\mathrm{mt}}^{\mathrm{hom}} = \frac{E^s}{1+\delta Q} \qquad \nu_{\mathrm{mt}}^{\mathrm{hom}} = \frac{\nu^s}{1+\delta Q} \tag{10.57}$$

which again should be compared to the homologous quantities derived in the dilute case (see equation [10.30]).

10.5.2. *Closed cracks*

In order to deal with the case of closed frictionless cracks ($\mathbb{C}^i = 3k^s\mathbb{J}$), it is readily seen from [10.50] that it suffices to replace $\mathbb{T}$ by $\mathbb{T}'$ in [10.52]:

$$3\mathrm{D}: \mathbb{C}_{\mathrm{mt}}^{\mathrm{hom}} = \mathbb{C}^s : \left(\mathbb{I} + \frac{4\pi}{3|\Omega|} a_i^3 \mathbb{T}_i'\right)^{-1} \qquad 2\mathrm{D}: \mathbb{C}_{\mathrm{mt}}^{\mathrm{hom}} = \mathbb{C}^s : \left(\mathbb{I} + \frac{\pi}{|S_\Omega|} a_i^2 \mathbb{T}_i'\right)^{-1} \tag{10.58}$$

Accordingly, for parallel cracks, $\mathbb{T}$ is replaced by $\mathbb{T}'$ in [10.53]

$$\begin{aligned} 3\text{D} : \mathbb{C}_{\text{mt}}^{\text{hom}} &= \mathbb{C}^s : \left(\mathbb{I} + \frac{4\pi d}{3}\mathbb{T}'\right)^{-1} \\ 2\text{D} : \mathbb{C}_{\text{mt}}^{\text{hom}} &= \mathbb{C}^s : \left(\mathbb{I} + \pi\delta\mathbb{T}'\right)^{-1} \end{aligned} \qquad [10.59]$$

Similarly, for randomly oriented cracks, $< \mathbb{T} >$ is replaced by $< \mathbb{T}' >$ (see equations [10.38] and [10.43]) in [10.54].

In the 3D case, it is readily seen that:

$$k_{\text{mt}}^{\text{hom}} = k^s; \quad \mu_{\text{mt}}^{\text{hom}} = \frac{\mu^s}{1 + Q'_\mu d} \qquad [10.60]$$

with Q'_μ given in [10.40]. The estimates can be compared to the dilute counterparts (see equation [10.39]) as well as to [9.16] and [9.20] (Griffith crack model). The corresponding expressions of the Young's modulus and Poisson ratio are:

$$\begin{aligned} E_{\text{mt}}^{\text{hom}} &= \frac{E^s}{1 + Q'_E d} \\ \nu_{\text{mt}}^{\text{hom}} &= (1 + \nu^s)\frac{1 + Q'_\mu d}{1 + Q'_E d} - 1 \end{aligned} \qquad [10.61]$$

with Q'_E given in [10.42].

In the 2D case, the Young's modulus and Poisson ratio in the plane perpendicular to the direction $\underline{e}_3$ in which the cracks are infinite read:

$$\begin{aligned} E_{\text{mt}}^{\text{hom}} &= \frac{E^s}{1 + \delta Q/4} \\ \nu_{\text{mt}}^{\text{hom}} &= \frac{\nu^s + \delta Q/4}{1 + \delta Q/4} \end{aligned} \qquad [10.62]$$

with Q given in [10.30]. These expressions are to be compared to the dilute corresponding quantities (see equation [10.44]).

11

Stress-based Estimates of Stress Concentration and Compliance

11.1. Dilute estimate of the stress concentration tensor

The damaging effect related to the presence of the cracks is represented in [8.36] (since $\varphi_i \to 0$) by the term $\varphi_i \mathbb{S}_i^c : \overline{\mathbb{B}}^i$. We therefore need to estimate the average stress rate in a crack. The approach presented in this section again belongs to the Eshelby-based estimates in the so-called dilute framework. It consists of estimating the stresses in a given crack by use of the ones which take place when the crack is embedded in an infinite medium with compliance $\mathbb{S}^s$, subjected to a uniform stress at infinity equal to the macroscopic stress $\boldsymbol{\Sigma} : \boldsymbol{\sigma}(\underline{z}) \to \boldsymbol{\Sigma}$ when $|\underline{z}| \to \infty$. This amounts to stating that the strain $\boldsymbol{\varepsilon}(\underline{z})$ tends toward $\mathbb{S}^s : \boldsymbol{\Sigma}$ at infinity. We then retrieve the classical (strain-based) boundary conditions of the Eshelby problem, which provides the average strain $\overline{\boldsymbol{\varepsilon}}^c$ in the crack. The average stress in the crack thus reads $\overline{\boldsymbol{\sigma}}^c = \mathbb{C}^c : \overline{\boldsymbol{\varepsilon}}^c$ from which the following dilute estimate of $\overline{\mathbb{B}}^c$ is derived:

$$\overline{\mathbb{B}}^c_{\text{dil}} = \mathbb{C}^c : \left(\mathbb{I} + \mathbb{P} : (\mathbb{C}^c - \mathbb{C}^s)\right)^{-1} : \mathbb{S}^s \qquad [11.1]$$

Let us introduce the symmetric fourth-order tensor $\mathbb{Q}$ defined as

$$\mathbb{Q} = \mathbb{C}^s - \mathbb{C}^s : \mathbb{P} : \mathbb{C}^s = \mathbb{C}^s : (\mathbb{I} - \mathbb{S}) \qquad [11.2]$$

from which we can see that:

$$\mathbb{C}^s : \mathbb{P} = \mathbb{I} - \mathbb{Q} : \mathbb{S}^s \qquad [11.3]$$

With this notation, $\overline{\mathbb{B}}^c_{dil}$ also reads

$$\overline{\mathbb{B}}^c_{\text{dil}} = \mathbb{C}^c : \left(\mathbb{C}^s + \mathbb{C}^s : \mathbb{P} : (\mathbb{C}^c - \mathbb{C}^s)\right)^{-1} = \mathbb{C}^c : \left(\mathbb{Q} + \mathbb{C}^s : \mathbb{P} : \mathbb{C}^c\right)^{-1} \qquad [11.4]$$

[11.3] is then introduced into the right-hand side of the above equation:

$$\overline{\mathbb{B}}^c_{\text{dil}} = \mathbb{C}^c : \left(\mathbb{Q} + \mathbb{C}^c - \mathbb{Q} : \mathbb{S}^s : \mathbb{C}^c\right)^{-1} = \left(\mathbb{I} + \mathbb{Q} : (\mathbb{S}^c - \mathbb{S}^s)\right)^{-1} \qquad [11.5]$$

Introducing the above expression of the stress concentration tensor into [8.36] provides the stress-based dilute estimate of the effective compliance tensor:

$$\mathbb{S}^{\text{hom}}_{\text{dil}} = \mathbb{S}^s + \sum_i \varphi_i \left((\mathbb{S}^c_i - \mathbb{S}^s)^{-1} + \mathbb{Q}_i\right)^{-1} \qquad [11.6]$$

11.2. Dilute estimates of the effective compliance for opened cracks

The case of opened cracks is characterized by the condition $\mathbb{S}^c_i \to \infty$; [11.6] then produces

$$\mathbb{S}^{\text{hom}}_{\text{dil}} = \mathbb{S}^s + \sum_i \varphi_i \mathbb{Q}_i^{-1} \qquad [11.7]$$

which can be clearly defined if we consider opened penny-shaped cracks with radius a_i. The volume fraction of the elementary crack $\mathcal{C}_i$ is again written as $\varphi_i = X_i\varphi'_i$, where φ'_i is equal to $4\pi a_i^3/(3|\Omega|)$ in the 3D case and to $\pi a_i^2/|\mathcal{S}_\Omega|$ in the 2D case. Since $(\mathbb{I} - \mathbb{S}_i)^{-1}$ is of the order $1/X_i$, $\mathbb{Q}_i^{-1} = (\mathbb{I} - \mathbb{S}_i)^{-1} : \mathbb{S}^s$ is singular when the aspect ratio $X_i \to 0$ but there exists a limit for $X_i\mathbb{Q}_i^{-1}$, as well as for $\varphi_i\mathbb{Q}_i^{-1}$. Hence, the stress-based estimate $\mathbb{S}_{\text{dil}}^{\text{hom}}$ derived for opened cracks from [11.7] when X_i and $\varphi_i \to 0$ reads:

$$\begin{pmatrix} 3\text{D} : \mathbb{S}_{\text{dil}}^{\text{hom}} = \left(\mathbb{I} + \dfrac{4\pi}{3|\Omega|}a_i^3\mathbb{T}_i\right) : \mathbb{S}^s \\ 2\text{D} : \mathbb{S}_{\text{dil}}^{\text{hom}} = \left(\mathbb{I} + \dfrac{\pi}{|\mathcal{S}_\Omega|}a_i^2\mathbb{T}_i\right) : \mathbb{S}^s \end{pmatrix} \qquad [11.8]$$

which proves to be the inverse to the stiffness tensor $\mathbb{C}_{\text{mt}}^{\text{hom}}$ provided by the strain-based Mori–Tanaka scheme [WIL 80, DOR 06] derived in [10.52]. However, is important to emphasize that the latter differs from the strain-based dilute estimate of the effective stiffness (see equation [10.16]).

In the particular cases of opened parallel cracks or of opened randomly oriented cracks, the stress-based dilute estimate of the effective compliance can now be obtained from the inverse of the stiffness in [10.53] and [10.54].

11.2.1. *Opened parallel cracks*

$$\begin{aligned} 3\text{D} : \mathbb{S}_{\text{dil}}^{\text{hom},3D} &= \left(\mathbb{I} + \frac{4\pi d}{3}\mathbb{T}\right) : \mathbb{S}^s \\ 2\text{D} : \mathbb{S}_{\text{dil}}^{\text{hom},2D} &= \left(\mathbb{I} + \pi\delta\mathbb{T}\right) : \mathbb{S}^s \end{aligned} \qquad [11.9]$$

With regard to the expression of the 2D stiffness tensor of [11.9], the question is again: should we use the 2D determination $\mathbb{T} = \mathbb{T}^{ps}$ or the 3D determination $\mathbb{T} = \mathbb{T}^{cyl}$ (see equation [10.15])? The answer is similar to that provided in the context of strain-based estimates in section 10.3. In short, the matrices of $\mathbb{S}_{\text{dil}}^{\text{hom},2D}$ obtained with either choice are indeed different. Moreover, the choice $\mathbb{T} = \mathbb{T}^{ps}$ yields a non-symmetric tensor $\mathbb{S}_{\text{dil}}^{\text{hom},2D}$. Nevertheless, the restrictions of both tensors to the vector subspace of stress tensors $\mathbf{\Sigma}$ that meet the conditions of plane strain:

$$\left(\mathbb{S}_{\text{dil},2D}^{\text{hom}} : \mathbf{\Sigma}\right) : \underline{e}_i \overset{s}{\otimes} \underline{e}_3 = 0 \qquad [11.10]$$

are identical. Again, this validates the use of the 2D determination of $\mathbb{T}$, namely $\mathbb{T} = \mathbb{T}^{ps}$.

However, we can see that the vector subspace defined by [11.10] a priori itself depends on the choice of $\mathbb{T}$. Based on [10.13]–[10.15], it is evident that [11.10] in fact reduces to the conditions of plane strain for the sane (uncracked) material, irrespective of whether $\mathbb{T} = \mathbb{T}^{cyl}$ or $\mathbb{T} = \mathbb{T}^{ps}$:

$$(\mathbb{S}^s : \mathbf{\Sigma}) : \underline{e}_i \overset{s}{\otimes} \underline{e}_3 = 0 \qquad [11.11]$$

which can be put in the form

$$\Sigma_{13} = \Sigma_{23} = 0; \quad \Sigma_{33} = \nu(\Sigma_{11} + \Sigma_{22}) \qquad [11.12]$$

and we can verify that $\left(\mathbb{I} + \pi\delta\mathbb{T}^{ps}\right) : \mathbb{S}^s$ is indeed a symmetric tensor on the subspace defined by [11.12], in the sense that:

$$\mathbf{\Sigma} : \left(\mathbb{I} + \pi\delta\mathbb{T}^{ps}\right) : \mathbb{S}^s : \mathbf{\Sigma}' = \mathbf{\Sigma}' : \left(\mathbb{I} + \pi\delta\mathbb{T}^{ps}\right) : \mathbb{S}^s : \mathbf{\Sigma} \qquad [11.13]$$

11.2.2. *Opened randomly oriented cracks*

For opened randomly oriented cracks, [11.8] takes the form:

$$\begin{aligned} 3\text{D}: \mathbb{S}_{\text{dil}}^{\text{hom}} &= \Big(\mathbb{I} + \frac{4\pi d}{3} < \mathbb{T} > \Big) : \mathbb{S}^s \\ 2\text{D}: \mathbb{S}_{\text{dil}}^{\text{hom}} &= \Big(\mathbb{I} + \pi\delta < \mathbb{T} > \Big) : \mathbb{S}^s \end{aligned} \qquad [11.14]$$

11.2.3. *Discussion*

As already stated, the dilute estimate of the effective stiffness $\mathbb{C}_{\text{dil}}^{\text{hom}}$ derived in the framework of uniform strain boundary conditions at infinity, and the dilute estimate of the effective compliance $\mathbb{S}_{\text{dil}}^{\text{hom}}$ derived in the framework of uniform stress boundary conditions at infinity are not inverse of one another:

$$\mathbb{C}_{\text{dil}}^{\text{hom}} : \mathbb{S}_{\text{dil}}^{\text{hom}} \neq \mathbb{I} \qquad [11.15]$$

This raises the question of the consistency of this homogenization scheme.

However, it is evident, for instance in the case of parallel cracks, from a combination of [10.19]–[10.21] and [11.9] that $\mathbb{C}_{\text{dil}}^{\text{hom}} : \mathbb{S}_{\text{dil}}^{\text{hom}} \to \mathbb{I}$ when $d \to 0$ (respectively, $\delta \to 0$). More precisely:

$$\mathbb{C}_{\text{dil}}^{\text{hom}} : \mathbb{S}_{\text{dil}}^{\text{hom}} = \mathbb{I} + \mathcal{O}(d^2) \quad (\text{resp. } \mathbb{I} + \mathcal{O}(\delta^2)) \qquad [11.16]$$

This emphasizes the fact that the relevance of the dilute scheme is not controlled by the crack volume fraction (which is always infinitesimal) but by the crack density parameter d (respectively, δ).

11.3. Dilute estimate of the effective compliance for closed cracks

Again following [DEU 02], a closed crack is represented by a flat ellipsoid filled with an isotropic linear elastic material with bulk and shear moduli k^c and μ^c:

$$\mathbb{S}^c = \frac{1}{3k^c}\mathbb{J} + \frac{1}{2\mu^c}\mathbb{K} \qquad [11.17]$$

Furthermore, the closed frictionless crack only transfers normal stresses and therefore has a vanishing shear modulus: $\mu^c \to 0$. In order to remain consistent with the model adopted in the approach based on uniform strain boundary conditions [10.31], it is required that $k^c = k^s$. Accordingly, we have:

$$\mathbb{S}^c - \mathbb{S}^s = m\mathbb{K} \quad \text{with} \quad m = \frac{1}{2\mu^c} - \frac{1}{2\mu^s} \to \infty \qquad [11.18]$$

Note that [11.6] cannot be used directly since the quantity $(\mathbb{S}^c - \mathbb{S}^s)^{-1}$ is not defined. Therefore, introducing [11.18] into [11.5] and [8.36] gives the following dilute estimate of the homogenized compliance:

$$\mathbb{S}^{\text{hom}}_{\text{dil}} = \mathbb{S}^s + \sum_i m\varphi_i\mathbb{K} : \left(\mathbb{I} + m\mathbb{Q}_i : \mathbb{K}\right)^{-1} \qquad [11.19]$$

We now introduce $b = 1/m$, noting that

$$\mathbb{S}^{\text{hom}}_{\text{dil}} = \mathbb{S}^s + \sum_i \varphi_i\mathbb{K} : \left(b\mathbb{I} + \mathbb{Q}_i : \mathbb{K}\right)^{-1} \quad \text{with} \quad b \to 0 \qquad [11.20]$$

It should be observed that it is not possible to take the limit $b \to 0$ in the above equation by replacing b with 0 in $\left(b\mathbb{I} + \mathbb{Q} : \mathbb{K}\right)^{-1}$. The difficulty comes from the fact that $\left(\mathbb{Q} : \mathbb{K}\right)^{-1}$

is not defined. The idea is to introduce the quantity $\mathcal{B}(\beta,b)=\left(\beta\mathbb{J}+b\mathbb{K}+\mathbb{Q}:\mathbb{K}\right)^{-1}$, which is obviously equal to $\left(b\mathbb{I}+\mathbb{Q}:\mathbb{K}\right)^{-1}$ for $b=\beta$. The interesting feature[1] is that the product $\mathbb{K}:\mathcal{B}(\beta,b)$ is in fact independant of β, *provided that* $\beta\neq 0$ (recall that the tensor $\mathcal{B}(\beta,b)$ is not defined for $\beta=0$). It follows that any value of $\beta\neq 0$ can be chosen:

$$(\forall\beta\neq 0)\quad \mathbb{K}:\left(b\mathbb{I}+\mathbb{Q}:\mathbb{K}\right)^{-1}=\mathbb{K}:\mathcal{B}(\beta,b) \qquad [11.21]$$

For a given *fixed* value of β, we now note that $\mathbb{K}:\mathcal{B}(\beta,b)$ is a smooth function of b. This implies that:

$$(\forall\beta\neq 0)\lim_{b\to 0}\mathbb{K}:\mathcal{B}(\beta,b)=\mathbb{K}:\left(\beta\mathbb{J}+\mathbb{Q}:\mathbb{K}\right)^{-1}=\mathbb{K}:\mathcal{B}(\beta,0)[11.22]$$

Combining [11.21] and [11.22] we find that, irrespective of the value of $\beta\neq 0$:

$$\lim_{b\to 0}\mathbb{K}:\left(b\mathbb{I}+\mathbb{Q}:\mathbb{K}\right)^{-1}=\mathbb{K}:\mathcal{B}(\beta,0) \qquad [11.23]$$

11.3.1. *3D case*

In the 3D case, the tensor $\mathbb{K}:\mathcal{B}(\beta,0)$ in the right-hand side of [11.23] can be determined in the framework of the Walpole algebra:

$$3D:\quad \lim_{b\to 0}\mathbb{K}:\left(b\mathbb{I}+\mathbb{Q}:\mathbb{K}\right)^{-1}=\frac{2(1-\nu)}{\mu(2-\nu)\pi X}\mathbb{E}^{(4)}+\mathcal{O}(X^0)\,[11.24]$$

in which[2] $\mathbb{E}^{(4)}=\boldsymbol{i}_N\overline{\underline{\otimes}}\boldsymbol{i}_T+\boldsymbol{i}_T\overline{\underline{\otimes}}\boldsymbol{i}_N$ with $\underline{n}$ being the unit normal to the considered crack, $\boldsymbol{i}_N=\underline{n}\otimes\underline{n}$ and $\boldsymbol{i}_T=\mathbf{1}-\underline{n}\otimes\underline{n}$. Then,

1 The proof of this statement can be derived in the framework of the tensorial algebra introduced by Walpole (see section 11.5).

2 For the definition of the $\overline{\underline{\otimes}}$ operator, see section 11.5.

[11.24] is introduced into [11.20]:

$$\begin{aligned} \mathbb{S}^{\text{hom}}_{\text{dil}} &= \mathbb{S}^s + \frac{1}{|\Omega|}\frac{8(1-\nu)}{3(2-\nu)\mu}\sum_i a_i^3 \mathbb{E}_i^{(4)} \\ &= \left(\mathbb{I} + \frac{16(1-\nu)}{3(2-\nu)}\frac{1}{|\Omega|}\sum_i a_i^3 \mathbb{E}_i^{(4)}\right) : \mathbb{S}^s \end{aligned} \qquad [11.25]$$

In an orthonormal frame $(\underline{t}_1, \underline{t}_2, \underline{n})$, all coefficients $E^{(4)}_{ijkl}$ are equal to 0 except

$$E^{(4)}_{2323} = E^{(4)}_{1313} = 1/2 \qquad [11.26]$$

In agreement with the physical expectation, [11.25] states that a closed frictionless crack with normal $\underline{n}$ introduces additional compliance in a macroscopic shear loading $\Sigma_{nt} \neq 0$.

11.3.2. *2D case*

The derivation is based on the expression of the coefficient of the Hill tensor for an elliptic inclusion (see equation [2.105]). In an orthonormal frame $(\underline{e}_1, \underline{e}_2, \underline{e}_3)$ with $\underline{e}_1 = \underline{t}$ and $\underline{e}_2 = \underline{n}$, it is evident that

$$Q_{1212} = \frac{E^s}{1-\nu^{s2}}\frac{X}{2(1+X^2)} \qquad [11.27]$$

from which it follows that the series expansion of $\mathbb{K} : \mathcal{B}(\beta, 0)$ reads:

$$\mathbb{K} : \mathcal{B}(\beta, 0) = \frac{1-\nu^s}{\mu^s X}\mathbb{E}^{(1212)} + \mathcal{O}(X^0) \qquad [11.28]$$

with the notations

$$\mathbb{E}^{(1212)} = \left(\underline{e}_1 \overset{s}{\otimes} \underline{e}_2\right) \otimes \left(\underline{e}_1 \overset{s}{\otimes} \underline{e}_2\right) \qquad [11.29]$$

and

$$\left(\underline{e}_1 \overset{s}{\otimes} \underline{e}_2\right) = \frac{1}{2}\left(\underline{e}_1 \otimes \underline{e}_2 + \underline{e}_2 \otimes \underline{e}_1\right) \qquad [11.30]$$

Note that the component $E_{1212}^{(1212)} = 1/4$; [11.28] is then introduced into [11.23], which in turn is used in [11.20]:

$$\begin{aligned} \mathbb{S}_{\text{dil}}^{\text{hom}} &= \mathbb{S}^s + \frac{\pi}{|\mathcal{S}_\Omega|} \frac{1-\nu^s}{\mu^s} \sum_i a_i^2 \mathbb{E}_i^{(1212)} \\ &= \left(\mathbb{I} + \frac{2\pi(1-\nu^s)}{|\mathcal{S}_\Omega|} \sum_i a_i^2 \mathbb{E}_i^{(1212)}\right) : \mathbb{S}^s \end{aligned} \qquad [11.31]$$

11.3.3. *Stress concentration tensor*

Interestingly, [11.1] also gives the "dilute" stress concentration tensor in the crack modeled as a 3D heterogeneity. The asymptotic value of $\overline{\mathbb{B}}^c$ when $m \to \infty$ reads:

$$\overline{\mathbb{B}}_{dil}^c = \lim_{m \to \infty} \left(\mathbb{I} + m\mathbb{Q} : \mathbb{K}\right)^{-1} = \mathbf{1} \otimes \underline{n} \otimes \underline{n} \qquad [11.32]$$

In the framework of 3D modeling of the crack, [11.32] implies that the average stress in the crack is the spherical tensor $\overline{\mathbb{B}}_{dil}^c : \mathbf{\Sigma} = \Sigma_{nn}\mathbf{1}$. This is consistent with frictionless crack lips, which only transfer normal stresses.

11.3.4. *Comparison with other estimates*

In order to compare [11.25] and [11.31] with the effective stiffness derived in [10.33] and in [10.58] in the framework of a strain-based approach, it is useful to recall the result presented in [11.23] in which the value $\beta = 3k^s$ is chosen. Recalling that $\mathbb{Q} = \mathbb{C}^s : (\mathbb{I} - \mathbb{S})$, the limit considered in [11.23] takes the form:

$$\begin{aligned} \lim_{b\to 0} \mathbb{K} : \left(b\mathbb{I} + \mathbb{Q} : \mathbb{K}\right)^{-1} &= \mathbb{K} : \left(3k^s\mathbb{J} + \mathbb{Q} : \mathbb{K}\right)^{-1} \\ &= \mathbb{K} : \left(\mathbb{C}^s - \mathbb{C}^s : \mathbb{S} : \mathbb{K}\right)^{-1} \\ &= \mathbb{K} : \left(\mathbb{I} - \mathbb{S} : \mathbb{K}\right)^{-1} : \mathbb{S}^s \end{aligned} \qquad [11.33]$$

We eventually introduce [11.33] into [11.20]:

$$\mathbb{S}_{\text{dil}}^{\text{hom}} = \left(\mathbb{I} + \sum_i \varphi_i' X_i \mathbb{K} : \left(\mathbb{I} - \mathbb{S}_i : \mathbb{K}\right)^{-1}\right) : \mathbb{S}^s \qquad [11.34]$$

with $\varphi_i' = 4\pi a_i^3/(3|\Omega|)$ in the 3D case (respectively, $\varphi_i' = \pi a_i^2/(|\mathcal{S}_\Omega|)$ in the 2D case). Hence, an expression involving the tensor $\mathbb{T}'$ introduced in [10.34] is obtained asymptotically as $X \to 0$:

$$\begin{aligned} 3D : \mathbb{S}_{\text{dil}}^{\text{hom}} &= \left(\mathbb{I} + \sum_i \frac{4\pi}{3|\Omega|} a_i^3 \mathbb{T}_i'\right) : \mathbb{S}^s \\ 2D : \mathbb{S}_{\text{dil}}^{\text{hom}} &= \left(\mathbb{I} + \sum_i \frac{\pi}{|\mathcal{S}_\Omega|} a_i^2 \mathbb{T}_i'\right) : \mathbb{S}^s \end{aligned} \qquad [11.35]$$

Using this form, it appears that the above (stress-based) dilute estimates of the effective compliance are the inverses of the (strain-based) Mori–Tanaka estimates of the effective stiffness (see equation [10.58]).

11.4. Mori–Tanaka estimates of effective compliance

In the stress-based approach, the Mori–Tanaka scheme consists of introducing an auxiliary stress tensor Σ_o and replacing the remote boundary conditions by $\boldsymbol{\sigma}(\underline{z}) \to \boldsymbol{\Sigma}_o$ when $|\underline{z}| \to \infty$. The average stress in the crack thus reads $\overline{\boldsymbol{\sigma}}^c = \mathbb{C}^c : \left(\mathbb{I} + \mathbb{P} : (\mathbb{C}^c - \mathbb{C}^s)\right)^{-1} : \mathbb{S}^s : \boldsymbol{\Sigma}_o$, while the average stress in the matrix is $\overline{\boldsymbol{\sigma}}^s = \boldsymbol{\Sigma}_o$. The average rule $\overline{\boldsymbol{\sigma}} = \boldsymbol{\Sigma}$ provides the link between $\boldsymbol{\Sigma}_o$ and $\boldsymbol{\Sigma}$ from which the following expressions of the stress concentration tensors are derived:

$$\overline{\mathbb{B}}^c_{mt} = \left(\mathbb{I} + \mathbb{Q} : (\mathbb{S}^c - \mathbb{S}^s)\right)^{-1} : \Big((1-\varphi)\mathbb{I} + \sum_i \varphi_i \left(\mathbb{I} + \mathbb{Q}_i : (\mathbb{S}^c_i - \mathbb{S}^s)\right)^{-1}\Big)^{-1} \quad [11.36]$$

and

$$\overline{\mathbb{B}}^s_{mt} = \Big((1-\varphi)\mathbb{I} + \sum_i \varphi_i \left(\mathbb{I} + \mathbb{Q}_i : (\mathbb{S}^c_i - \mathbb{S}^s)\right)^{-1}\Big)^{-1} \quad [11.37]$$

Eventually, [11.36] is introduced into [8.36] and provides the Mori–Tanaka estimate of effective compliance:

$$\mathbb{S}^{\text{hom}}_{\text{mt}} = \mathbb{S}^s + \sum_i \varphi_i \left((\mathbb{S}^c_i - \mathbb{S}^s)^{-1} + \mathbb{Q}_i\right)^{-1} : \Big((1-\varphi)\mathbb{I} + \sum_i \varphi_i \left(\mathbb{I} + \mathbb{Q}_i : (\mathbb{S}^c_i - \mathbb{S}^s)\right)^{-1}\Big)^{-1} \quad [11.38]$$

where $\varphi = \sum_i \varphi_i$. We note that the expressions of $\mathbb{S}^{\text{hom}}_{\text{dil}}$ and $\mathbb{S}^{\text{hom}}_{\text{mt}}$ differ by a coefficient, which is equal to $\overline{\mathbb{B}}^s_{mt}$:

$$\mathbb{S}^{\text{hom}}_{\text{mt}} - \mathbb{S}^s = \left(\mathbb{S}^{\text{hom}}_{\text{dil}} - \mathbb{S}^s\right) : \overline{\mathbb{B}}^s_{mt} \quad [11.39]$$

11.4.1. *Opened cracks*

We again use the fact that opened cracks are characterized by the condition $\mathbb{S}_i^c \to \infty$; [11.38] then returns

$$\mathbb{S}_{\mathrm{mt}}^{\mathrm{hom}} = \mathbb{S}^s + \frac{1}{1-\varphi}\sum_i \varphi_i \mathbb{Q}_i^{-1} \qquad [11.40]$$

Owing to the fact that the crack porosity $\varphi \to 0$, the stress-based dilute estimate derived in [11.7] and the above Mori–Tanaka estimate then prove to be identical for cracked media. Note that the estimates given in [11.7] and [11.40] remain valid for any porous medium and that, in general, the added compliance due to the porosity in the stress-based dilute and Mori–Tanaka estimates differ by a factor $1/(1-\varphi)$.

11.4.2. *Closed cracks*

Comparing [11.5] and [11.37], it appears that the average stress concentration $\overline{\mathbb{B}}_{mt}^s$ also reads:

$$\overline{\mathbb{B}}_{mt}^s = \left((1-\varphi)\mathbb{I} + \sum_i \varphi_i \overline{\mathbb{B}}_{dil,i}^c\right)^{-1} \qquad [11.41]$$

We recall that $\overline{\mathbb{B}}_{dil,i}^c$ has a finite limit as $m \to \infty$ (see equation [11.32]). It follows that $\overline{\mathbb{B}}_{mt}^s = \mathbb{I}$ asymptotically, since φ as well as all $\varphi_i \to 0$; [11.39] implies that the dilute and Mori–Tanaka estimates of effective compliance are identical for closed cracks. This extends the result already mentioned for opened cracks.

11.5. Appendix: algebra for transverse isotropy and applications

The homogenized properties of the medium which consists of an isotropic solid matrix with parallel cracks exhibit transversely isotropic symmetry. For this reason, we introduce standard notations and the corresponding simplified algebra [WAL 81] for fourth-order transverse isotropic tensors. This algebra is particularly useful to carry out easily inner product of transversely isotropic tensors or their inversion. Given a unit vector $\underline{n}$, we introduce $\boldsymbol{i}_N = \underline{n} \otimes \underline{n}$ and $\boldsymbol{i}_T = 1 - \underline{n} \otimes \underline{n}$, as well as:

$$\begin{aligned} \mathbb{E}_1 &= \frac{1}{2} \boldsymbol{i}_T \otimes \boldsymbol{i}_T; \; \mathbb{E}_2 = \boldsymbol{i}_N \otimes \boldsymbol{i}_N; \mathbb{E}_3 = \boldsymbol{i}_T \overline{\underline{\otimes}} \boldsymbol{i}_T - \mathbb{E}_1; \\ \mathbb{E}_4 &= \boldsymbol{i}_N \overline{\underline{\otimes}} \boldsymbol{i}_T + \boldsymbol{i}_T \overline{\underline{\otimes}} \boldsymbol{i}_N \end{aligned} \qquad [11.42]$$

with the following definition of the $\overline{\underline{\otimes}}$ operator:

$$(\boldsymbol{a} \overline{\underline{\otimes}} \boldsymbol{b})_{ijkl} = \frac{1}{2} \left(a_{ik} b_{jl} + a_{il} b_{jk} \right) \qquad [11.43]$$

It is evident that:

$$\mathbb{E}_1 + \mathbb{E}_2 + \mathbb{E}_3 + \mathbb{E}_4 = \mathbb{I} \qquad [11.44]$$

and

$$\mathbb{E}_p : \mathbb{E}_q = \mathbb{E}_p \quad \text{if } p = q \; ; \; \mathbb{E}_p : \mathbb{E}_q = 0 \quad \text{if } p \neq q; \qquad [11.45]$$

Two other elementary tensors are also defined:

$$\mathbb{E}_5 = \boldsymbol{i}_N \otimes \boldsymbol{i}_T \; ; \; \mathbb{E}_6 = \boldsymbol{i}_T \otimes \boldsymbol{i}_N \; ; \qquad [11.46]$$

From these identities, we show that any transversely isotropic (not necessarily symmetric) fourth-order tensor can be decomposed as:

$$\mathbb{L} = c\mathbb{E}_1 + d\mathbb{E}_2 + e\mathbb{E}_3 + f\mathbb{E}_4 + g\mathbb{E}_5 + h\mathbb{E}_6 \qquad [11.47]$$

which may be expressed in the symbolic form:

$$\mathbb{L} = [c, d, e, f, g, h] \qquad [11.48]$$

For instance, the tensors $\mathbb{I}$, $\mathbb{J}$ and $\mathbb{K}$ are represented by:

$$\mathbb{I} = [1, 1, 1, 1, 0, 0] \quad \mathbb{J} = [\frac{2}{3}, \frac{1}{3}, 0, 0, \frac{1}{3}, \frac{1}{3}]$$

$$\mathbb{K} = [\frac{1}{3}, \frac{2}{3}, 1, 1, -\frac{1}{3}, -\frac{1}{3}] \qquad [11.49]$$

Moreover, in the isotropic case, the elastic stiffness and compliance are represented by:

$$\mathbb{C}^s = \frac{E^s}{(1+\nu)(1-2\nu)}[1, 1-\nu, 1-2\nu, 1-2\nu, \nu, \nu]$$

$$\mathbb{S}^s = \frac{1}{E^s}[1-\nu, 1, 1+\nu, 1+\nu, -\nu, -\nu] \qquad [11.50]$$

If $\mathbb{L}$ is symmetric ($\mathbb{L} = {}^t\mathbb{L}$), as for stiffness or compliance tensor, then $g = h$.

With this notation, considering $\mathbb{L}' = [c', d', e', f', g', h']$, the tensor product $\mathbb{L} : \mathbb{L}'$ reads:

$$\mathbb{L} : \mathbb{L}' = [cc' + 2hg', dd' + 2gh', ee', ff', gc' + dg', hd' + ch'] \qquad [11.51]$$

and the inverse of $\mathbb{L}$ is

$$\mathbb{L}^{-1} = [\frac{d}{l}, \frac{c}{l}, \frac{1}{e}, \frac{1}{f}, -\frac{g}{l}, -\frac{h}{l}] \qquad [11.52]$$

with $l = cd - 2gh$.

Note that the non-symmetric tensor $(\mathbb{I} - \mathbb{S}(X, \underline{n}))$ can be recast in the form $[1 - c, 1 - d, 1 - e, 1 - f, -g, -h]$ with, for $\underline{n} = \underline{e}_3$:

$$c = S_{1111} + S_{1122}; \quad d = S_{3333}; \quad e = S_{1111} - S_{1122}; \qquad [11.53]$$
$$f = 2S_{3131}; \quad g = S_{3311}; \quad h = S_{1133}$$

Recall that the Eshelby tensor is related to the Hill tensor by $\mathbb{S} = \mathbb{P} : \mathbb{C}^s$. The components of P_{ijkl} are defined in [2.126] for $\underline{n} = \underline{e}_3$.

12

Bounds

In Chapters 10 and 11, we proposed various Eshelby-based estimates of the effective stiffness tensor $\mathbb{C}^{\mathrm{hom}}$ of a microcracked medium. The purpose of this chapter is to present the strain-based variational approach of homogenization[1]. The goal is to derive exact upper bounds of $\mathbb{C}^{\mathrm{hom}}$. This alternative approach will be detailed in the 3D case. For the most part, the developments of this chapter are general and go beyond the application to cracked materials.

The problem of heterogeneous elasticity being considered is summarized by the following equations:

$$\begin{array}{lll} \operatorname{div}\boldsymbol{\sigma} = 0 & (\Omega) & (a) \\ \boldsymbol{\sigma} = \mathbb{C}(\underline{z}) : \boldsymbol{\varepsilon} & (\Omega) & (b) \\ \boldsymbol{\varepsilon} = \dfrac{1}{2}(\nabla\underline{\xi} + {}^{t}\nabla\underline{\xi}) & (\Omega) & (c) \\ \underline{\xi}(\underline{z}) = \mathbf{E}\cdot\underline{z} & (\partial\Omega) & (d) \end{array} \qquad [12.1]$$

in which Ω is the considered *r.e.v.* and the unknown displacement, strain and stress fields defined on Ω are denoted by $\underline{\xi}$, $\boldsymbol{\varepsilon}$ and $\boldsymbol{\sigma}$. The macroscopic strain tensor $\mathbf{E}$ is the loading parameter.

---

1 For a stress-based approach, see also [DOR 09],[WIL 81],[WIL 97]

12.1. The energy definition of the homogenized stiffness

The purpose of this section is to provide an alternative definition of the homogenized stiffness tensor $\mathbb{C}^{\text{hom}}$ based on an energy approach. This new definition will prove to be equivalent to the direct approach stated in [8.26].

For the time being, we will disregard the question of geometrical nonlinearity that was raised in section 8.2.3. The starting point is to express the elastic energy W stored in the *r.e.v.* Ω at the macroscopic scale when it is subjected to the macroscopic strain $\mathbf{E}$. In the framework of linear elasticity, we have:

$$W = \frac{|\Omega|}{2}\mathbf{E} : \mathbb{C}^{\text{hom}} : \mathbf{E} \qquad [12.2]$$

We now consider the mechanical power provided to the *r.e.v.* Ω

$$\dot{W} = \int_{\Omega} \sigma : \dot{\varepsilon}\, dV_z = \int_{\Omega} \varepsilon : \mathbb{C} : \dot{\varepsilon}\, dV_z \qquad [12.3]$$

where the local state equation $\sigma = \mathbb{C} : \varepsilon$ was used. Integration with respect to time returns:

$$\dot{W} = \int_{\Omega} \varepsilon : \mathbb{C} : \dot{\varepsilon}\, dV_z \Rightarrow W = \frac{1}{2}\int_{\Omega} \varepsilon : \mathbb{C} : \varepsilon\, dV_z \qquad [12.4]$$

We then introduce the strain concentration rule [8.25] into [12.4]:

$$W = \frac{1}{2}\mathbf{E} : \int_{\Omega} {}^{t}\mathbb{A} : \mathbb{C} : \mathbb{A}\, dV_z : \mathbf{E} \qquad [12.5]$$

A comparison of [12.5] and [12.2] returns:

$$\mathbb{C}^{\text{hom}} = \overline{{}^{t}\mathbb{A} : \mathbb{C} : \mathbb{A}} \qquad [12.6]$$

Attention must be paid to the apparent difference between expressions [8.26] and [12.6] of the homogenized tensor of stiffness. The consistency of the two definitions can be established with the help of the Hill lemma. More precisely, we can apply the fourth-order tensor $\overline{{}^t\mathbb{A} : \mathbb{C} : \mathbb{A}}$ to the couple $(\mathbf{E}', \mathbf{E}'')$ of two symmetric tensors:

$$\mathbf{E}' : \overline{{}^t\mathbb{A} : \mathbb{C} : \mathbb{A}} : \mathbf{E}'' = \overline{\varepsilon' : \mathbb{C} : \varepsilon''} \tag{12.7}$$

where $\varepsilon' = \mathbb{A} : \mathbf{E}'$ (resp. $\varepsilon'' = \mathbb{A} : \mathbf{E}''$) is the microscopic strain field solution of the elasticity problem [12.1] associated with $\mathbf{E}'$ (respectively, $\mathbf{E}''$). This implies that:

1) ε' is a geometrically compatible strain field that complies with a uniform strain boundary condition and with the average rule $\overline{\varepsilon'} = \mathbf{E}'$;

2) $\sigma'' = \mathbb{C} : \varepsilon''$ satisfies the momentum balance equation $\operatorname{div} \sigma'' = 0$. Its average is $\overline{\mathbb{C} : \mathbb{A}} : \mathbf{E}''$.

Hill's lemma [8.18] applied to the couple (σ'', ε') returns:

$$\mathbf{E}' : \overline{{}^t\mathbb{A} : \mathbb{C} : \mathbb{A}} : \mathbf{E}'' = \overline{\varepsilon' : \sigma''} = \underbrace{\mathbf{E}'}_{\overline{\varepsilon'}} : \underbrace{\overline{\mathbb{C} : \mathbb{A}} : \mathbf{E}''}_{\overline{\sigma''}} \tag{12.8}$$

The consistency of [8.26] and [12.6] is a consequence of the fact that the above identity holds for an arbitrary choice of the couple $(\mathbf{E}', \mathbf{E}'')$:

$$\overline{{}^t\mathbb{A} : \mathbb{C} : \mathbb{A}} = \overline{\mathbb{C} : \mathbb{A}} \tag{12.9}$$

Note that the energy definition allows us to establish the symmetry of the homogenized stiffness tensor in its direct form [8.26].

An important conclusion of this section is combining the micro- and the macroscopic approaches to defining the elastic energy stored in the *r.e.v.* :

$$W = \frac{|\Omega|}{2}\mathbf{E} : \mathbb{C}^{\text{hom}} : \mathbf{E} = \frac{1}{2}\int_\Omega \varepsilon : \mathbb{C} : \varepsilon \, dV_z \qquad [12.10]$$

For forthcoming use in Chapter 13, it is useful to introduce the volume density $\Psi = W/|\Omega|$ of elastic energy at the macroscopic scale:

$$\Psi = \frac{1}{2}\mathbf{E} : \mathbb{C}^{\text{hom}} : \mathbf{E} \qquad [12.11]$$

In [12.10], ε denotes the strain field associated with the displacement field $\underline{\xi}$ solution to problem [12.1]. The principle of minimum potential energy applied to the *r.e.v.* Ω subjected to the loading defined by $\mathbf{E}$ in the sense of the boundary condition [8.3] provides a variational characterization of $\underline{\xi}$ and ε. It states that $\underline{\xi}$ minimizes the potential energy among all kinematically admissible displacements[2] $\underline{\xi}'$. Since in the present problem the potential energy is identical to the elastic energy (no given forces), we have:

$$(\forall \text{ k.a. } \underline{\xi}') \quad \frac{|\Omega|}{2}\mathbf{E} : \mathbb{C}^{\text{hom}} : \mathbf{E} = \frac{1}{2}\int_\Omega \varepsilon : \mathbb{C} : \varepsilon \, dV_z$$
$$\leq \frac{1}{2}\int_\Omega \varepsilon'(\underline{\xi}') : \mathbb{C} : \varepsilon'(\underline{\xi}') \, dV_z \qquad [12.12]$$

A straightforward application of this principle of minimum is the so-called Voigt bound on $\mathbb{C}^{\text{hom}}$. We simply apply the above condition to the field $\underline{\xi}' = \mathbf{E} \cdot \underline{z}$, which is obviously k.a.

2 That is, meeting the boundary condition [8.3]:

$$(\forall \underline{z} \in \partial\Omega) \quad \underline{\xi}'(\underline{z}) = \mathbf{E} \cdot \underline{z}$$

with E. Since the corresponding strain ε' is uniform and equal to E, [12.12] takes the form:

$$\mathbf{E} : \mathbb{C}^{\text{hom}} : \mathbf{E} \leq \mathbf{E} : \overline{\mathbb{C}} : \mathbf{E} \quad \Leftrightarrow \quad \mathbb{C}^{\text{hom}} \leq \overline{\mathbb{C}} \qquad [12.13]$$

For example, let us consider a cracked material. As before, let $\varphi = \sum_i \varphi_i$ denote the total crack porosity. The Voigt bound states that $\mathbb{C}^{\text{hom}} \leq (1-\varphi)\overline{\mathbb{C}}^s$. In the case of cracks, the crack porosity being negligible, the Voigt bound [12.13] thus reduces[3] to $\mathbb{C}^{\text{hom}} \leq \overline{\mathbb{C}}^s$.

From a rigorous viewpoint, the above developments are only valid if the geometrical nonlinearity is negligible. In fact, as discussed in section 8.2.3, the concept of strain concentration tensor and the associated assumption of linearity may become questionable for infinitesimal crack aspect ratios. The range of macroscopic loading for which the elastic energy can be put in the form [12.5] is accordingly reduced. This restriction should be kept in mind when expressions like [12.11] are used.

12.2. Hashin–Shtrikman's bound

12.2.1. *Hashin–Shtrikman variational principle*

The present section is inspired by [WIL 77]. Consider a numerical fonction $v(\varepsilon')$ of the tensorial variable ε' (symmetric second-order tensor). The Legendre transform is defined by:

$$v^*(\tau) = \inf_{\varepsilon'} \left(\tau : \varepsilon' - v(\varepsilon')\right) \qquad [12.14]$$

3 For a homogeneous solid phase, note that $\mathbb{C}^s = \overline{\mathbb{C}}^s$. The Voigt bound fails to account for damaging effects of cracks.

When $v(\varepsilon')$ is a definite negative quadratic form $\frac{1}{2}\varepsilon' : \mathbb{C} : \varepsilon'$, it is evident that $v^*(\tau)$ is finite and given by $\frac{1}{2}\tau : \mathbb{C}^{-1} : \tau$.
Let Ω now denote a *r.e.v.* of a composite. At the point $\underline{z}$ of Ω (microscopic scale), the local elasticity tensor is $\mathbb{C}(\underline{z})$. We introduce a linear elastic fictitious medium (elasticity tensor $\mathbb{C}_o$), referred to as a "reference medium", such that:

$$(\forall \underline{z} \in \Omega) \quad \mathbb{C}_o > \mathbb{C}(\underline{z}) \qquad [12.15]$$

meaning that $\mathbb{C}(\underline{z}) - \mathbb{C}_o$ is negative definite, in the sense of quadratic forms. For instance, in a porous material with homogeneous solid phase with stiffness tensor $\mathbb{C}^s$, a possible choice is $\mathbb{C}_o = \lambda \mathbb{C}^s$ with $\lambda > 1$.

The corresponding elastic energy density is denoted by $w_o(\varepsilon') = \frac{1}{2}\varepsilon' : \mathbb{C}_o : \varepsilon'$. For later use, we also need to consider the associated Green operator Γ_o as defined in section 2.4.3. The subscript o in Γ_o refers to the stiffness tensor $\mathbb{C}_o$, which replaces $\mathbb{C}$ in the polarization problem [2.59]. From the definition of $(w - w_o)^*(\tau)$, we obtain:

$$(\forall \varepsilon', \tau) \quad w(\varepsilon') \leq w_o(\varepsilon') + \tau : \varepsilon' - (w - w_o)^*(\tau) \qquad [12.16]$$

The minimum principle [12.12] can be put in the form[4]:

$$\Psi(\mathbf{E}) = \frac{1}{2}\mathbf{E} : \mathbb{C}^{\text{hom}} : \mathbf{E} = \frac{1}{|\Omega|} \inf_{\varepsilon' \, k.a. \, \mathbf{E}} \int_\Omega w(\varepsilon') \, dV \qquad [12.17]$$

4 The kinematically admissible displacement fields are characterized by the boundary condition [12.1d].

Consider now an arbitrary polarization field $\boldsymbol{\tau}$ defined on Ω. By integration of [12.16] over the *r.e.v.* , this returns:

$$(\forall\tau)\ \Psi(\mathbf{E}) \leq \frac{1}{|\Omega|}\left(\inf_{\varepsilon'\ c.a.\ \mathbf{E}} \int_\Omega w_o(\varepsilon') + \tau : \varepsilon'\, dV \right.$$
$$\left. - \int_\Omega (w - w_o)^*(\tau)\, dV \right) \qquad [12.18]$$

For the polarization field being considered, the minimum of the first integral in the right-hand side is reached in the solution of the polarization problem [2.59], which is given by [2.62]: $\varepsilon' = \mathbf{E} - \Gamma_o * \tau$ is therefore introduced into [12.18] :

$$(\forall\tau)\ \Psi(\mathbf{E})\ \leq \frac{1}{|\Omega|}\left(\int_\Omega w_o(\mathbf{E} - \Gamma_o * \tau) + \tau : (\mathbf{E} \right.$$
$$\left. -\Gamma_o * \tau)\, dV - \int_\Omega (w - w_o)^*(\tau)\, dV \right) \qquad [12.19]$$

Observing that $\overline{\Gamma_o * \boldsymbol{\tau}} = 0$ (see equation [2.63]), we observe that:

$$\int_\Omega w_o(\mathbf{E} - \Gamma_o * \tau)\, dV = \frac{|\Omega|}{2}\mathbf{E} : \mathbb{C}_o : \mathbf{E}$$
$$+\frac{1}{2}\int_\Omega \Gamma_o * \tau : \mathbb{C}_o : \Gamma_o * \tau\, dV \qquad [12.20]$$

By definition of the Green operator, $\tau - \mathbb{C}_o : (\Gamma_o * \tau)$ can be interpreted as the stress field, which is the solution to the polarization problem [2.59] with $\mathbf{E} = 0$. It therefore meets the condition $\operatorname{div}\left(\tau - \mathbb{C}_o : (\Gamma_o * \tau)\right) = 0$. Again using $\overline{\Gamma_o * \boldsymbol{\tau}} = 0$, Hill's lemma [8.18] with $\sigma = \tau - \mathbb{C}_o : (\Gamma_o * \tau)$ and $\varepsilon = -\Gamma_o * \tau$ reads:

$$\int_\Omega \Gamma_o * \tau : (\tau - \mathbb{C}_o : \Gamma_o * \tau)\ dV = 0 \qquad [12.21]$$

or

$$\int_{\Omega} \Gamma_o * \tau : \mathbb{C}_o : \Gamma_o * \tau \, dV = \int_{\Omega} \tau : \Gamma_o * \tau \, dV \qquad [12.22]$$

With these results, [12.19] becomes:

$$(\forall \tau) \quad \Psi(\mathbf{E}) \leq w_o(\mathbf{E}) + \overline{\tau} : \mathbf{E} - \frac{1}{2}\overline{\tau : \Gamma_o * \tau} - \frac{1}{2}\overline{\tau : (\mathbb{C} - \mathbb{C}_o)^{-1} : \tau} \qquad [12.23]$$

It is convenient to put [12.23] in the form:

$$(\forall \tau) \quad \frac{1}{2}\mathbf{E} : \mathbb{C}^{\mathrm{hom}} : \mathbf{E} \leq \frac{1}{2}\mathbf{E} : \mathbb{C}_o : \mathbf{E} + \mathcal{H}(\tau) \qquad [12.24]$$

where $\mathcal{H}(\tau)$ is a quadratic functional of the polarization field referred to as the Hashin functional:

$$\mathcal{H}(\tau) = \overline{\tau} : \mathbf{E} - \frac{1}{2}\overline{\tau : \Gamma_o * \tau} - \frac{1}{2}\overline{\tau : (\mathbb{C} - \mathbb{C}_o)^{-1} : \tau} \qquad [12.25]$$

Through [12.24], any choice of a polarization field provides an upper bound of $\mathbb{C}^{\mathrm{hom}}$. For instance, consider a porous material with porosity φ. If the solid is homogeneous, let the reference medium be defined by $\mathbb{C}_o = \lambda \mathbb{C}^s$. The particular choice $\tau = 0$ yields $\mathbb{C}^{\mathrm{hom}} \leq \lambda \mathbb{C}^s$. We therefore retrieve the bound $\mathbb{C}^{\mathrm{hom}} \leq \mathbb{C}^s$ (with $\lambda \to 1^+$). Alternatively, let the polarization field τ be uniform, equal to some tensor τ_o. In this case, we have $\Gamma_o * \tau = 0$ and the optimal choice for τ_o is found to be:

$$\tau_o = \overline{(\mathbb{C} - \mathbb{C}_o)^{-1}}^{-1} : \mathbf{E} \qquad [12.26]$$

The corresponding bound is:

$$\mathbb{C}^{\mathrm{hom}} \leq \mathbb{C}_o + \overline{(\mathbb{C} - \mathbb{C}_o)^{-1}}^{-1} \qquad [12.27]$$

The last step consists of selecting the optimal reference medium. Considering again $\mathbb{C}_o = \lambda\mathbb{C}^s$, the best upper bound is found for $\lambda \to \infty$ and is none other than the Voigt bound: $\mathbb{C}^{\text{hom}} \leq (1-\varphi)\mathbb{C}^s$.

12.2.2. *Piecewise constant polarization field*

In this section, we assume that Ω is made up of 2 homogeneous phases Ω^i ($i = s$ and c); respectively, the homogeneous solid (stiffness tensor $\mathbb{C}^s$) and the cracks. The cracks are assumed to be opened and are represented by an elastic medium with vanishing stiffness ($\mathbb{C}^c \to 0$) (see [8.23]). For any value of $\lambda > 1$, the reference medium with stiffness $\mathbb{C}_o = \lambda\mathbb{C}^s$ obviously meets the condition [12.15]. In particular, we consider the asymptotic choice $\lambda \to 1^+$, which entails using the solid itself as the reference medium, so that from now on $\mathbb{C}_o = \mathbb{C}^s$. However, in order for the last term in $\mathcal{H}(\tau)$ not to go to infinity, we have to prescribe the condition $\boldsymbol{\tau}_{|\Omega^s} = 0$. Among the possible polarization fields, we explore piecewise constant fields, which are completely characterized by a given polarization tensor $\boldsymbol{\tau}_c$ for the crack domain Ω^c:

$$\tau(\underline{z}) = \boldsymbol{\tau}_c \chi^c(\underline{z}) \qquad [12.28]$$

where $\chi^c(\underline{z})$ is the indicator function of Ω^c. Recalling [2.65] for the derivation of $\overline{\tau : \Gamma_o * \tau}$, we obtain:

$$\mathcal{H}(\tau) = \varphi\tau_c : \mathbf{E} - \frac{1}{2}\varphi\tau_c : (\mathbb{C}_c - \mathbb{C}_o)^{-1} : \tau_c - \frac{1}{2}\tau_c : \mathbb{A}_{cc} : \tau_c \qquad [12.29]$$

where $\mathbb{A}_{cc}$ is given by:

$$\mathbb{A}_{cc} = \frac{1}{|\Omega|}\int_\Omega \chi^c(\underline{z})\Gamma_o^\infty(\underline{z} - \underline{z}')(\chi^c(\underline{z}') - \varphi)\, dV_z dV_{z'} \qquad [12.30]$$

As seen from [2.51], the symmetry of $\mathbb{\Gamma}_o^\infty(\underline{z}-\underline{z}')$ implies that $\mathbb{A}_{cc}$ is itself a fourth-order symmetric tensor, irrespective of the geometry of Ω^c.

When the geometry of the microstructure of Ω is known in a deterministic way, a numerical computation of $\mathbb{A}_{cc}$ is feasible, at least theoretically. The optimal choice τ_c^{opt} of τ_c is the one that minimizes $\mathcal{H}(\tau)$. This returns:

$$\varphi\mathbf{E} = \left(\varphi\left(\mathbb{C}^c - \mathbb{C}^s\right)^{-1} + \mathbb{A}_{cc}\right) : \tau_c^{\text{opt}} \qquad [12.31]$$

Considering the limit $\mathbb{C}^c \to 0$, [12.31] becomes

$$\varphi\boldsymbol{\tau}_c^{\text{opt}} = -\mathbb{C}^s : \mathbb{D} : E \quad \text{with} \quad \mathbb{D} = \varphi\left(\mathbb{I} - \frac{1}{\varphi}\mathbb{A}_{cc} : \mathbb{C}^s\right)^{-1} \qquad [12.32]$$

It follows that[5]:

$$\frac{1}{2}\mathbf{E} : \mathbb{C}^{\text{hom}} : \mathbf{E} \leq \frac{1}{2}\mathbf{E} : \mathbb{C}^s : \mathbf{E} + \frac{\varphi}{2}\tau_c^{\text{opt}} : \mathbf{E} \qquad [12.33]$$

which can be put in the form of an upper bound of $\mathbb{C}^{\text{hom}}$:

$$\mathbb{C}^{\text{hom}} \leq \mathbb{C}^s : (\mathbb{I} - \mathbb{D}) \qquad [12.34]$$

5 Consider a second-order polynomial function of the form:

$$\mathcal{P}(\boldsymbol{\tau}) = \mathbf{a} : \boldsymbol{\tau} + \frac{1}{2}\boldsymbol{\tau} : \mathbb{B} : \boldsymbol{\tau}$$

where $\mathbb{B}$ is a symmetric, inversible tensor. The extremum reached for $\boldsymbol{\tau}^{opt} = -\mathbb{B}^{-1} : \mathbf{a}$ is $-\frac{1}{2}\mathbf{a} : \mathbb{B}^{-1} : \mathbf{a}$, that is $\frac{1}{2}\mathbf{a} : \boldsymbol{\tau}^{opt}$.

For later use, using the expression of $\mathbb{D}$ given in [12.32], note that the upper bound in [12.34] also reads:

$$\mathbb{C}_s : (\mathbb{I} - \mathbb{D}) = \mathbb{C}_s : \left((1-\varphi)\mathbb{I} - \frac{1}{\varphi}\mathbb{A}_{cc} : \mathbb{C}_s\right) : \left(\mathbb{I} - \frac{1}{\varphi}\mathbb{A}_{cc} : \mathbb{C}_s\right)^{-1} \qquad [12.35]$$

$\mathbb{D}$ can be interpreted as a macroscopic (fourth-order) damage tensor, which incorporates information concerning both the elastic properties of the solid and the geometry of the crack network[6]. In general, $\mathbb{D}$ is not symmetric. In contrast, the symmetry of $\mathbb{A}_{cc}$ ensures that of $\mathbb{C}^s : \mathbb{D}$. Indeed, observing that

$$\mathbb{D} = \varphi \mathbb{S}^s : \left(\mathbb{S}^s - \frac{1}{\varphi}\mathbb{A}_{cc}\right)^{-1} \qquad [12.36]$$

yields

$$\mathbb{C}^s : \mathbb{D} = \varphi \left(\mathbb{S}^s - \frac{1}{\varphi}\mathbb{A}_{cc}\right)^{-1} \qquad [12.37]$$

12.2.3. *Random microstructures*

Let us briefly mention a first point of view that consists of the random generation of realizations of the microstructure geometry of the *r.e.v.*, formally denoted by ω, the volume fraction φ being given.

6 The limit $\varphi \to 0$ (case of cracks) in [12.32] must be taken with care and will be detailed later.

The tensor $\mathbb{A}_{cc}$ of [12.30] depends on the particular realization ω according to:

$$\mathbb{A}_{cc}(\omega) = \frac{1}{|\Omega|} \int_{\Omega} \chi^c(\omega, \underline{z}) \boldsymbol{\Gamma}_o^{\infty}(\underline{z} - \underline{z}')(\chi^c(\omega, \underline{z}') - \varphi) \, dV_z dV_{z'} \quad [12.38]$$

In turn, the tensor $\mathbb{D}$ of [12.32] depends on ω through $\mathbb{A}_{cc}(\omega)$ and the corresponding upper bound reads:

$$\mathbb{C}^{\text{hom}}(\omega) \leq \mathbb{C}_s : (\mathbb{I} - \mathbb{D}(\omega)) \quad [12.39]$$

For practical use, it is appealing to consider the average value of this upper bound over the realization that is denoted by brackets ($< \cdot >$):

$$< \mathbb{C}^{\text{hom}}(\omega) > \leq \mathbb{C}^s : (\mathbb{I} - < \mathbb{D}(\omega) >) \quad [12.40]$$

The outlook described herein is the one adopted in the celebrated Hashin–Shtrikman bounds [HAS 63]. It consists of seeking a polarization tensor (τ_c^{opt}), which should be optimal in a statistical sense. This means that it is expected to be the optimum when a statistical set of realizations of the *r.e.v.* is considered. The polarization tensor (τ_c^{opt}) becomes deterministic insofar as it does not refer to any particular realization of the *r.e.v.*. Practically, this amounts to replacing $\mathcal{H}(\omega, \tau)$ in [12.24] by its statistical average denoted by $< \mathcal{H} > (\tau)$:

$$< \mathcal{H} > (\tau) = \varphi \tau_c : \mathbf{E} - \frac{1}{2} \varphi \tau_c : (\mathbb{C}_c - \mathbb{C}_o)^{-1} : \tau_c - \frac{1}{2} \tau_c :< \mathbb{A}_{cc} >: \tau_c \quad [12.41]$$

So far, we have not distingushed the crack domain from the pore space of an arbitrary porous material. In order to determine $< \mathbb{A}_{cc} >$, we introduce appropriate morphological and statistical assumptions concerning the cracks.

First, it is assumed that they are identical in shape and orientation. Their shape is represented by an ellipsoid I. Let $\chi(\underline{z})$ denote the characteristic of the latter, centered at the origin of the space. Hence, the characteristic function of crack $\mathcal{C}_\alpha$ centered at $\underline{z}_\alpha$ reads $\chi_\alpha(\underline{z}) = \chi(\underline{z} - \underline{z}_\alpha)$. It is useful to introduce the norm defined by $\mathbf{q}^I$:

$$|\underline{z}|_I = (\underline{z} \cdot \mathbf{q}^I \cdot \underline{z})^{1/2} \qquad [12.42]$$

The cartesian equation of ellipsoid I is:

$$\mathcal{Q}(\underline{z}) = |\underline{z}|_I - 1 \leq 0 \qquad [12.43]$$

By definition of the characteristic function $\chi(\underline{z})$, the following conditions are equivalent:

$$\begin{array}{lll} \chi(\underline{z}) = 1 & \Leftrightarrow & \mathcal{Q}(\underline{z}) \leq 0 \\ \chi(\underline{z}) = 0 & \Leftrightarrow & \mathcal{Q}(\underline{z}) > 0 \end{array} \qquad [12.44]$$

More generally, let I_λ denote the ellipsoid obtained from I by the homothety with ratio λ and center O. Its equation is:

$$\mathcal{Q}_\lambda(\underline{z}) = |\underline{z}|_I - \lambda \leq 0 \qquad [12.45]$$

Following [PON 95], let us now introduce the statistical assumption on which this approach is based: it states that the spatial distribution of cracks follows an ellipsoidal symmetry defined by a positive definite second-order tensor q^D. The latter is *a priori* different from the tensor q^I of [12.43], which characterizes the shape of the cracks. More precisely, consider a given crack center at $\underline{z}_\alpha$ and let $f(\underline{r})$ denote the probability density function such that $f(\underline{r})dV_r$ represents the number of cracks centered at $\underline{z}_\beta$ for which $\underline{r}_{\alpha\beta} = \underline{z}_\beta - \underline{z}_\alpha$ belongs to the elementary sphere dV_r centered at $\underline{r}$. The ellipsoidal symmetry assumption states that $f(\underline{r})$ depends on $\underline{r}$ through the quadratic quantity $(\underline{r} \cdot \mathbf{q}^D \cdot \underline{r})^{1/2} = |\underline{r}|_D$:

$$f(\underline{r}) = F\left(|\underline{r}|_D\right) \qquad [12.46]$$

The tensor $\mathbf{q}^D$ allows us to introduce a family of ellipsoids D_λ defined as in [12.45]:

$$\underline{z} \in D_\lambda \Leftrightarrow |\underline{z}|_D - \lambda \leq 0 \qquad [12.47]$$

In the following, it will prove convenient to represent the *r.e.v.* Ω as an ellipsoid D_λ with $\lambda \to \infty$. The characteristic function of the crack domain takes the form:

$$\chi^c(\underline{z}) = \sum_\alpha \chi_\alpha(\underline{z}) \qquad [12.48]$$

Let us introduce this expression into [12.38]:

$$\mathbb{A}_{cc}(\omega) = \frac{1}{|\Omega|} \int_\Omega \mathbb{\Gamma}_o^\infty(\underline{z} - \underline{z}') \sum_\alpha \chi(\underline{z} - \underline{z}_\alpha) \times \left(\sum_\beta \chi(\underline{z}' - \underline{z}_\beta) - \varphi\right) dV_z dV_{z'} \qquad [12.49]$$

which is conveniently split into two terms

$$\mathbb{A}_{cc}(\omega) = \mathbb{A}_{cc}^{(1)} - \mathbb{A}_{cc}^{(2)} \qquad [12.50]$$

where:

$$\mathbb{A}_{cc}^{(1)} = \frac{1}{|\Omega|} \sum_{\alpha,\beta} \int_\Omega \mathbb{\Gamma}_o^\infty(\underline{z} - \underline{z}')\chi(\underline{z} - \underline{z}_\alpha)\chi(\underline{z}' - \underline{z}_\beta)\, dV_z dV_{z'} \qquad [12.51]$$

$$\mathbb{A}_{cc}^{(2)} = \frac{\varphi}{|\Omega|} \sum_{\alpha} \int_{\Omega} \Gamma_o^{\infty}(\underline{z} - \underline{z}')\chi(\underline{z} - \underline{z}_\alpha)\, dV_z dV_{z'} \qquad [12.52]$$

We shall begin with the second term that reads:

$$\mathbb{A}_{cc}^{(2)} = \frac{\varphi}{|\Omega|} \sum_{\alpha} \int_{\mathcal{C}_\alpha} dV_z \int_{\Omega} \Gamma_o^{\infty}(\underline{z} - \underline{z}')\, dV_{z'} \qquad [12.53]$$

Recalling [2.57] with $\Omega = D_\lambda$, its evident that:

$$\mathbb{A}_{cc}^{(2)} = \varphi^2 \mathbb{P}^D \qquad [12.54]$$

We now consider $\mathbb{A}_{cc}^{(1)}$, which is put in the form

$$\mathbb{A}_{cc}^{(1)} = \frac{1}{|\Omega|} \sum_{\alpha} (\sum_{\beta} J_{\alpha\beta}) \qquad [12.55]$$

with

$$J_{\alpha\beta} = \int_{\Omega} \Gamma_o^{\infty}(\underline{z} - \underline{z}')\chi(\underline{z} - \underline{z}_\alpha)\chi(\underline{z}' - \underline{z}_\beta)\, dV_z dV_{z'} \qquad [12.56]$$

Observing that, in a statistical sense, the sum $\sum_\beta J_{\alpha\beta}$ should be independent of the choice of α, [12.55] becomes

$$\mathbb{A}_{cc}^{(1)} = \frac{\varphi}{|\mathcal{P}_\alpha|} (J_{\alpha\alpha} + \sum_{\beta \neq \alpha} J_{\alpha\beta}) \qquad [12.57]$$

Using [2.57] once more, it is evident that

$$J_{\alpha\alpha} = |\mathcal{P}_\alpha| \mathbb{P}^I \qquad [12.58]$$

The change in variables $\underline{u} = \underline{z} - \underline{z}_\alpha$ and $\underline{u}' = \underline{z}' - \underline{z}_\beta$ in [12.56] yields:

$$\sum_{\beta \neq \alpha} J_{\alpha\beta} = \int \Big(\sum_{\beta \neq \alpha} \mathbf{\Gamma}_o^\infty(\underline{u} - \underline{u}' - \underline{r}_{\alpha\beta})\Big)\chi(\underline{u})\chi(\underline{u}')\, dV_z dV_{z'} \qquad [12.59]$$

with $\underline{r}_{\alpha\beta} = \underline{z}_\beta - \underline{z}_\alpha$.

At this stage, the idea is to determine the statistical average of the above quantity $\sum_{\beta \neq \alpha} J_{\alpha\beta}$. To do so, we take advantage of the probability density function $f(\underline{r})$. Owing to its definition:

$$< \sum_{\beta \neq \alpha} \mathbf{\Gamma}_o^\infty(\underline{u} - \underline{u}' - \underline{r}_{\alpha\beta}) >= \int \mathbf{\Gamma}_o^\infty(\underline{u} - \underline{u}' - \underline{r}) f(\underline{r})\, dV_r \qquad [12.60]$$

It is sufficient to consider the above quantity only when $\underline{u}$ and $\underline{u}' \in I$, since otherwise the product $\chi(\underline{u})\chi(\underline{u}')$ in [12.59] is equal to 0. Therefore, we have[7] $|\underline{u} - \underline{u}'|_I \leq 2$ so that $\underline{u} - \underline{u}' \in I_2$ (see equation [12.45]). Moreover, if $\beta \neq \alpha$, it is evident that $\underline{r}_{\alpha\beta} \notin I_2$. It follows that the restriction of $f(\underline{r})$ to I_2 is 0:

$$f_{|I_2} = 0 \qquad [12.61]$$

In other words, we have:

$$<\sum_{\beta \neq \alpha} \mathbf{\Gamma}_o^\infty(\underline{u} - \underline{u}' - \underline{r}_{\alpha\beta})>= \int_{|\underline{r}|_I > 2} \mathbf{\Gamma}_o^\infty(\underline{u} - \underline{u}' - \underline{r}) f(\underline{r})\, dV_r \qquad [12.62]$$

12.2.3.1. *The case* $\mathbf{q}^D = \mathbf{q}^I$

It is instructive to first examine the case $\mathbf{q}^D = \mathbf{q}^I$. This means that the spatial distribution and the shape of the

7 By application of the classical triangular inequality.

cracks are governed by the same ellipsoidal symmetry. It is therefore possible to determine the integral[8] of [12.62] by successive integrations over layers $\mathcal{I}_{\lambda}^{\lambda+d\lambda}$ of infinitesimal thickness, located between the two ellipsoids I_λ and $I_{\lambda+d\lambda}$, with $\lambda > 2$. This entails evaluating the following integral:

$$\int_{\mathcal{I}_{\lambda}^{\lambda+d\lambda}} \mathbf{\Gamma}_o^{\infty}(\underline{u} - \underline{u}' - \underline{r}) f(\underline{r})\, dV_r \qquad [12.63]$$

The starting point is to observe that

$$(\forall \lambda > 2) \int_{I_\lambda} \mathbf{\Gamma}_o^{\infty}(\underline{u} - \underline{u}' - \underline{r})\, dV_r = \mathbb{P}_o^I \qquad [12.64]$$

which directly follows from [2.57], since $\underline{u} - \underline{u}' \in I_2$. This implies

$$\int_{\mathcal{I}_{\lambda}^{\lambda+d\lambda}} \mathbf{\Gamma}_o^{\infty}(\underline{u} - \underline{u}' - \underline{r})\, dV_r = \int_{I_{\lambda+d\lambda}} \mathbf{\Gamma}_o^{\infty}(\underline{u} - \underline{u}' - \underline{r})\, dV_r - \int_{I_\lambda} \mathbf{\Gamma}_o^{\infty}(\underline{u} - \underline{u}' - \underline{r})\, dV_r = 0 \qquad [12.65]$$

Let us go back to the quantity of [12.63] and remember the ellipsoidal symmetry [12.46] (with $D = I$). From [12.65], we obtain

$$\int_{\mathcal{I}_{\lambda}^{\lambda+d\lambda}} \mathbf{\Gamma}_o^{\infty}(\underline{u} - \underline{u}' - \underline{r}) f_\alpha(\underline{r})\, dV_r = F(\lambda) \int_{\mathcal{I}_{\lambda}^{\lambda+d\lambda}} \mathbf{\Gamma}_o^{\infty}(\underline{u} - \underline{u}' - \underline{r})\, dV_r = 0 \qquad [12.66]$$

8 For this derivation to be correct, it is intuitively necessary that the shape of domain Ω be homothetic to I. Concerning this point, our heuristic reasoning lacks the rigor found in [PON 95].

This proves that the sum [12.60] is equal to 0. Accordingly, from [12.59] we get:

$$< \sum_{\beta \neq \alpha} J_{\alpha\beta} >= 0 \qquad [12.67]$$

It follows from [12.57] and [12.58] that $< \mathbb{A}_{cc}^{(1)} >= \varphi \mathbb{P}^I$. Eventually, from [12.50] and [12.54] we obtain:

$$\mathbf{q}^D = \mathbf{q}^I \quad \Rightarrow \quad < \mathbb{A}_{cc} >= \varphi(1-\varphi)\mathbb{P}^I \qquad [12.68]$$

If we replace $\mathbb{A}_{cc}$ in [12.35] by the determination of $< \mathbb{A}_{cc} >$ given by [12.68], we get:

$$\mathbb{C}^s : (\mathbb{I} - \mathbb{D}) = (1-\varphi)\mathbb{C}^s : \left(\mathbb{I} - \mathbb{P}^I : \mathbb{C}^s\right) : \left(\mathbb{I} - (1-\varphi)\mathbb{P}^I : \mathbb{C}^s\right)^{-1} \qquad [12.69]$$

which also reads:

$$\mathbb{C}^s : (\mathbb{I} - \mathbb{D}) = (1-\varphi)\mathbb{C}^s : \left(\left(\mathbb{I} - (1-\varphi)\mathbb{P}^I : \mathbb{C}^s\right) : \left(\mathbb{I} - \mathbb{P}^I : \mathbb{C}^s\right)^{-1}\right)^{-1} \qquad [12.70]$$

and eventually

$$\mathbb{C}^s : (\mathbb{I} - \mathbb{D}) = (1-\varphi)\mathbb{C}^s : \left((1-\varphi)\mathbb{I} + \varphi\left(\mathbb{I} - \mathbb{P}^I : \mathbb{C}^s\right)^{-1}\right)^{-1} \qquad [12.71]$$

Using this form, it is interesting to note that the Hashin–Willis upper bound thus obtained is identical to the Mori–Tanaka estimate. Indeed, in the particular case of identical parallel cracks (same elasticity tensor $\mathbb{C}^i = 0$, Hill tensor $\mathbb{P}^i = \mathbb{P}^I$ for all cracks, crack porosity $\varphi = \sum_i \varphi_i$), [10.50] appears to be identical to [12.71].

12.2.3.2. *Case* $\mathrm{q}^D \neq \mathrm{q}^I$

We now address the general case $\mathrm{q}^D \neq \mathrm{q}^I$. The property [12.61] is still valid. Then, we shall consider the smallest value λ_o such that $I_2 \subset D_{\lambda_o}$. It follows that I_2 and D_{λ_o} are tangent at the contact points. Accordingly, at these points we have $f = 0$. Moreover, the ellipsoidal symmetry [12.46] implies that the condition $f = 0$ holds true on the whole ellipsoid D_{λ_o} (and for any point located inside D_{λ_o}). For this reason, we just have to replace [12.62] by the similar identity:

$$< \sum_{\beta \neq \alpha} \mathbf{\Gamma}_o^{\infty}(\underline{u} - \underline{u}' - \underline{r}_{\alpha\beta}) >= \int_{|\underline{r}|_D > \lambda_o} \mathbf{\Gamma}_o^{\infty}(\underline{u} - \underline{u}' - \underline{r}) f(\underline{r})\, dV_r \quad [12.72]$$

The reasoning then follows in a very similar manner: we again determine the r.h.s in [12.72] by integration over successive ellipsoidal layers $D_\lambda^{\lambda+d\lambda}$ located between D_λ and $D_{\lambda+d\lambda}$ so that [12.67] still holds. Accordingly, [12.68] is replaced by

$$< \mathbb{A}_{cc} >= \varphi \mathbb{P}^I - \varphi^2 \mathbb{P}^D \quad [12.73]$$

It now remains to incorporate this expression into [12.35]. This yields the Ponte-Castaneda and Willis upper bound [PON 95]:

$$\mathbb{C}^s : (\mathbb{I} - \mathbb{D}) = \mathbb{C}^s : \left(\mathbb{I} - \varphi\left((\mathbb{I} - \mathbb{P}^I : \mathbb{C}^s) + \varphi \mathbb{P}^D : \mathbb{C}^s\right)^{-1}\right) \quad [12.74]$$

The purpose of [12.74] is to clarify the specific roles of the shape of the inhomogeneities and of their spatial distribution. Since we are interested in cracks, we shall account for the shape through the model of a flat spheroid.

12.2.4. *Application of the Ponte-Castaneda and Willis (PCW) bound to microcracked media*

In the previous section, identical cracks were considered. The crack porosity thus takes the form

$$\varphi = \frac{4\pi}{3}\mathcal{N}a^3 X = \frac{4\pi d}{3}X$$

with the same notations as in Chapter 10. Owing to [10.10], the limit when $X \to 0$ of [12.74] reads:

$$\mathbb{C}^s : (\mathbb{I} - \mathbb{D}) = \mathbb{C}^s : \left(\mathbb{I} - \frac{4\pi d}{3}\mathbb{T} : \left(\mathbb{I} + \frac{4\pi d}{3}\mathbb{P}^D : \mathbb{C}^s : \mathbb{T}\right)^{-1}\right) \quad [12.75]$$

We shall briefly mention two important particular cases for the case of an isotropic linear elastic solid with compression modulus k^s, shear modulus μ^s, Young modulus E^s, and Poisson coefficient ν^s.

12.2.4.1. *Spherical distribution of parallel cracks*

The normal to the crack is oriented along $\underline{e}_1$. The distribution of the cracks in space is isotropic. $\mathbb{P}^D$ is the Hill tensor $\mathbb{P}^{\text{sph}}$ of a sphere (see equation [2.53]). The axial Young's modulus E_{11} and longitudinal shear modulus μ_{13} are given by:

– opened cracks:

$$\begin{aligned}\frac{E_{11}}{E^s} &= 1 - \frac{240d(1-\nu^{s2})}{45+16d(7-15\nu^{s2})}\\ \frac{\mu_{13}}{\mu^s} &= 1 - \frac{240d(1-\nu^s)}{45(2-\nu^s)+32d(4-5\nu^s)}\end{aligned} \quad [12.76]$$

– closed cracks:

$$\begin{aligned} &\frac{E_{11}}{E^s} = 1 \\ &\frac{\mu_{13}}{\mu^s} = 1 - \frac{240d(1-\nu^s)}{45(2-\nu^s)+32d(4-5\nu^s)} \end{aligned} \quad [12.77]$$

12.2.4.2. *Isotropic distribution of orientations*

The distribution of crack orientations is isotropic. The distribution of the cracks in space is isotropic.

– For opened cracks: The following results are established in [PON 95].

$$\begin{aligned} &\frac{k_{pcw}}{k^s} = 1 - \frac{48d(1-\nu^{s2})}{27(1-2\nu^s)+16d(1+\nu^s)^2} \\ &\frac{\mu_{pcw}}{\mu^s} = 1 - \frac{480d(1-\nu^s)(5-\nu^s)}{675(2-\nu^s)+64d(4-5\nu^s)(5-\nu^s)} \end{aligned} \quad [12.78]$$

– For closed cracks:

$$\begin{aligned} &\frac{k_{pcw}}{k^s} = 1 \\ &\frac{\mu_{pcw}}{\mu^s} = 1 - \frac{480d(1-\nu^s)}{225(2-\nu^s)+64d(4-5\nu^s)} \end{aligned} \quad [12.79]$$

13

Micromechanics-based Damage Constitutive Law and Application

13.1. Formulation of damage constitutive law

The lesson that we can learn from the results of homogenization of 3D (respectively, 2D) microcracked media is that the crack density parameter $d = \mathcal{N}a^3$ (respectively, $d = \mathcal{N}a^2$) is the relevant parameter with regard to the elastic energy. It is therefore natural to consider d – instead of the crack surface (respectively, length ℓ) – as the internal variable governing the effect of damage propagation. This is the focus of this section. However, the description of the damage state by a single scalar obviously narrows the domain of application to situations involving a single family of identical parallel cracks or identical cracks with random orientations that are isotropically distributed. However, in the latter case, an isotropic loading must be considered. In order to overcome this difficulty, it will be necessary to introduce a finite set of crack density parameters (d_i) $(i=1,\ldots,N)$, each corresponding to a specific family of microcracks.

13.1.1. *Description of damage level by a single scalar variable*

For the sake of simplicity, let us start with the case of a single crack density parameter d. One important conclusion of the homogenization approach as presented in Chapters 9–11 is that the effective stiffness tensor $\mathbb{C}^{\text{hom}}$ is a function of d. Accordingly, the macroscopic volume density of free energy Ψ appears to be a function of two state variables, namely the macroscopic strain $\mathbf{E}$ and d:

$$\Psi(\mathbf{E}, d) = \frac{1}{2}\mathbf{E} : \mathbb{C}^{\text{hom}}(d) : \mathbf{E} \qquad [13.1]$$

Now consider the intrinsic dissipation (per unit volume) in the form:

$$\mathcal{D} = \boldsymbol{\Sigma} : \dot{\mathbf{E}} - \dot{\Psi} \qquad [13.2]$$

The above expression of Ψ introduced into $\mathcal{D}$ yields:

$$\mathcal{D} = \left(\boldsymbol{\Sigma} - \frac{\partial \Psi}{\partial \mathbf{E}}\right) : \dot{\mathbf{E}} - \frac{\partial \Psi}{\partial d} : \dot{d} \qquad [13.3]$$

The classical reasoning consists in considering evolutions of the macroscopic strain that correspond to reversible transformations ($\mathcal{D} = 0$), which means $\dot{d} = 0$. The existence of such evolutions implies that:

$$\boldsymbol{\Sigma} = \frac{\partial \Psi}{\partial \mathbf{E}} = \mathbb{C}^{hom}(d) : \mathbf{E} \qquad [13.4]$$

Relation [13.3] now takes the form:

$$\mathcal{D} = -\frac{\partial \Psi}{\partial d}\dot{d} \geq 0 \qquad [13.5]$$

The driving force of microcrack propagation (respectively, damage propagation) therefore reads:

$$\mathcal{Y}_d = -\frac{\partial \Psi}{\partial d} = -\frac{1}{2}\mathbf{E} : \frac{\partial \mathbb{C}^{\mathrm{hom}}}{\partial d} : \mathbf{E} \qquad [13.6]$$

Indeed, by analogy with the energy release rate $G(\mathcal{C}, \ell)$ in [7.15], which is the driving force of fracture propagation ($\dot{\ell} \geq 0$), we can consider $\mathcal{Y}_d$ to be the driving force of damage propagation associated with an increase in the crack density parameter ($\dot{d} \geq 0$). Furthermore, given this analogy, it is appealing to adopt a criterion of the form [7.16] to describe damage propagation:

$$\mathcal{Y}_d - \mathcal{Y}_c \leq 0; \; \dot{d} \geq 0; \; (\mathcal{Y}_d - \mathcal{Y}_c)\, \dot{d} = 0 \qquad [13.7]$$

where $\mathcal{Y}_c$ is the analogous of the fracture energy G_c.

While it is obvious that quantities $\mathcal{Y}_c$ and G_c are physically related, no attempt will be presented here to connect them. This remains an open subject of reserach. In contrast to the fracture energy G_c, $\mathcal{Y}_c$ is *a priori* not an intrinsic material property, but could depend on the scale of observation and the crack size, and hence on the loading history.

Equations [13.6] and [13.7] turn out to be a strain formulation of the damage criterion. Alternatively, a stress formulation is obtained from a combination of [13.6] with the state equation $\mathbf{\Sigma} = \mathbb{C}^{\mathrm{hom}} : \mathbf{E}$.

Using the fact that $\mathbb{S}^{\mathrm{hom}} : \mathbb{C}^{\mathrm{hom}} = \mathbb{I}$, it is evident that

$$\frac{\partial}{\partial d}(\mathbb{C}^{\mathrm{hom}}) = -\mathbb{C}^{\mathrm{hom}} : \frac{\partial}{\partial d}(\mathbb{S}^{\mathrm{hom}}) : \mathbb{C}^{\mathrm{hom}} \qquad [13.8]$$

Introducing the above result into [13.6] yields:

$$\mathcal{Y}_d = \frac{1}{2}\mathbf{\Sigma} : \mathbb{H} : \mathbf{\Sigma} \leq \mathcal{Y}_c \qquad [13.9]$$

with

$$\mathbb{H} = \frac{\partial}{\partial d}(\mathbb{S}^{\text{hom}}) \qquad [13.10]$$

13.1.2. *Extension to multiple cracks*

A straightforward extension to multiple cracks of different radii and density would require a statistical analysis. We now consider the situation where the crack network is represented by several crack families, each one associated with a crack density parameter d^i. The radius is assumed to be uniform in each family, but can be different from one family to the other. An immediate extension of the methodology presented in section 13.1.1 consists of introducing a damage criterion of the type [13.7] for each crack family. Note that $\mathcal{Y}_c$ could *a priori* be different for each crack family because of its dependence on the crack size. But for the sake of simplicity, this dependency is disregarded:

$$(\forall i) \quad \mathcal{Y}_{d^i}(\mathbf{E}, d^1, \ldots, d^n) - \mathcal{Y}_c \leq 0 \qquad [13.11]$$

where $\mathcal{Y}_{d^i}$ is the driving force of the damage growth $(\delta d^1, \ldots, \delta d^n) \geq 0$:

$$\mathcal{Y}_{d^i} = -\frac{\partial \Psi}{\partial d^i} = -\frac{1}{2}\mathbf{E} : \frac{\partial \mathbb{C}^{\text{hom}}}{\partial d^i} : \mathbf{E} \qquad [13.12]$$

The dependency of $\mathcal{Y}_{d^i}$ on the d^j $(j = 1, \ldots, n)$ is captured throught the derivative of $\mathbb{C}^{\text{hom}}$ with respect to d^i, which can be determined by an appropriate homogenization scheme accounting for the interaction of cracks. The damage criterion associated with crack family i then reads:

$$\mathcal{Y}_{d^i} = -\frac{1}{2}\mathbf{E} : \frac{\partial}{\partial d^i}(\mathbb{C}^{\text{hom}}) : \mathbf{E} \leq \mathcal{Y}_c \qquad [13.13]$$

Expression [13.13] turns out to be a strain formulation of the damage criterion. Alternatively, a stress formulation similar to [13.9] can be derived from [13.13]. The counterpart of [13.8] reads

$$\frac{\partial}{\partial d^i}(\mathbb{C}^{\text{hom}}) = -\mathbb{C}^{\text{hom}} : \frac{\partial}{\partial d^i}(\mathbb{S}^{\text{hom}}) : \mathbb{C}^{\text{hom}} \qquad [13.14]$$

Introducing the above result into [13.13] yields:

$$(\forall i) \quad \mathcal{Y}_{d^i} = \frac{1}{2}\mathbf{\Sigma} : \mathbb{H}^i : \mathbf{\Sigma} \leq \mathcal{Y}_c \qquad [13.15]$$

with

$$\mathbb{H}^i = \frac{\partial}{\partial d^i}(\mathbb{S}^{\text{hom}}) \qquad [13.16]$$

We see that the propagation of the ith crack family is described by an elliptical stress criterion. The orientation of the crack family is taken into account through the anisotropic tensor $\mathbb{H}^i$.

13.2. Some remarks concerning the loss of uniqueness of the mechanical response in relation to damage

By analogy to [7.17], we can attempt to derive a criterion for stable damage propagation by a similar reasoning. If the dependence of $\mathcal{Y}_c$ on d is first disregarded, the transposition of

Chapter 7 would suggest at first sight that stability is ensured if

$$\frac{\partial \mathcal{Y}_d}{\partial d}(\mathbf{E}, d) = -\frac{\partial^2 \Psi}{\partial d^2}(\mathbf{E}, d) < 0 \qquad [13.17]$$

This criterion could be readily generalized if $\mathcal{Y}_c$ depends on d with a similar reasoning. More precisely, let us assume that the condition [13.7] for damage propagation is reached for the loading level **E** and the damage parameter d:

$$\mathcal{Y}_d(\mathbf{E}, d) = \mathcal{Y}_c(d) \qquad [13.18]$$

The damage propagation would be said to be stable if $\mathcal{Y}_d(\mathbf{E}, d + \delta d) < \mathcal{Y}_c(d + \delta d)$, that is if

$$\frac{\partial \mathcal{Y}_d}{\partial d}(\mathbf{E}, d) < \mathcal{Y}_c'(d) \qquad [13.19]$$

which also reads:

$$\frac{\partial^2 \Psi}{\partial d^2}(\mathbf{E}, d) + \mathcal{Y}_c'(d) > 0 \qquad [13.20]$$

In fact, the transposition of the reasoning, in Chapter 7, requires that the loading level **E** is prescribed, which is only possible at the scale of the *r.e.v* for which uniform strain boundary conditions are written on the displacement. In contrast, at the scale of a macroscopic structure in which the *r.e.v* corresponds to a material point, the local strain **E** is no longer prescribed, and the condition [13.20] cannot be used for the damage stability analysis. In order to illustrate this remark, a one-dimensional (1D) example is now presented.

We consider a rectilinear bar with length L. At any time t, the (macroscopic) displacement ξ is $\xi(0, t) = 0$ at $x = 0$ and is $\xi(L, t) = \xi_L(t)$ at $x = L$. The 1D local behavior reads $\Sigma =$

$C(d)E$ where the local (macroscopic) strain is $E = \dfrac{\partial \xi}{\partial x}(x,t)$. The condition of damage propagation [13.7] here reads:

$$-\frac{1}{2}C'(d)E^2 = \mathcal{Y}_c(d) \qquad [13.21]$$

Let us seek a uniform solution to the problem. At any time, the deformation reads $E_u(t) = \xi_L(t)/L$. In turn, the damage $d_u(t)$ can be derived from [13.21] and the stress is $\Sigma_u(t) = C(d_u(t))E(t)$.

Let us consider some loading level where the condition for damage propagation is fulfilled. We assume that the current state of the structure is defined by the uniform fields Σ_u, E_u and d_u. Let $\delta\Sigma$, δE and δd denote a possibly non-uniform incremental evolution associated with the loading increment $\delta\xi_L$.

The pertubation of the state equation takes the form:

$$\delta\Sigma = C(d_u)\delta E + C'(d_u)E_u\delta d \qquad [13.22]$$

In turn, pertubation of the damage criterion reads:

$$-\mathcal{Y}_c'(d_u)\delta d = C'(d_u)E_u\delta E + \frac{1}{2}C''(d_u)E_u^2\delta d \qquad [13.23]$$

The combination of [13.22] and [13.23] yields:

$$\delta\Sigma = C_t(E_u,d_u)\delta E \quad \text{with} \quad C_t(E_u,d_u) = C(d_u) - \frac{2C'(d_u)^2E_u^2}{C''(d_u)E_u^2 + 2\mathcal{Y}_c'(d_u)} \qquad [13.24]$$

As long as $C_t(E_u,d_u) \neq 0$ the momentum balance condition $\frac{\partial}{\partial x}(\delta\Sigma) = 0$ in turn implies that $\partial\delta E/\partial x = 0$, meaning that the incremental strain must be uniform. The uniqueness of the uniform solution follows.

Otherwise, if $C_t(E_u, d_u) = 0$, the incremental stress is uniform but the incremental strain δE loses its uniqueness and stability. This shows that [13.20] is not a relevant stability criterion.

13.3. Mechanical fields and damage in a hollow sphere subjected to traction

13.3.1. *General features*

We investigate now the response of a hollow sphere with an elastic damageable matrix and subjected to a uniform radial stress T on its external boundary. The internal (respectively, external) radius is denoted by a_o (respectively, b). The boundary conditions read:

$$\begin{array}{l} r = a_o : \sigma_{rr} = 0 \\ r = b : \quad \sigma_{rr} = T \end{array} \qquad [13.25]$$

We would like to emphasize the fact that the material point which follows in this chapter corresponds to the *r.e.v.* worked on so far. In this respect, the strain and stress tensors denoted by ε and σ in what follows are related to the current material point. They correspond to the tensors previously denoted by $\mathbf{E}$ and Σ. However, the notation $\underline{\xi}$ is kept for the displacement.

For the sake of simplicity and in order to derive exact solutions in closed form, no attempt is made to account for a possible stress-induced anisotropic damage. The local constitutive behavior of the solid at the so-called mesoscopic scale[1] ($a_o < r < b$) is assumedly described by isotropic damage

1 The problem being considered three length scales: at the *microscopic* scale, the material is made up of a linear elastic solid and microcracks. At the *mesoscopic* scale, we have a homogenized solid with bulk and shear moduli k and $\mu(d)$. We seek the response of the hollow sphere as a whole, which can be referred to as the *macroscopic* behavior.

modeled by a single scalar parameter d. In fact, the isotropy of the damage corresponds to a first-order approximation of the real microcracks density parameter distribution. Obviously, a more complex representation of general microcracking states will require at least a second-order damage tensor (see, for instance, [KRA 96]).

The specific isotropic damage model considered for the following structural analysis is defined by a constant bulk modulus k and a shear modulus $\mu(d)$ depending on a scalar damage variable d:

$$\mathbb{C}(d) = 3k\mathbb{J} + 2\mu(d)\mathbb{K} \quad [13.26]$$

This type of elastic damage model was proposed in a phenomenological framework by [DES 94] in the context of a discussion on uniqueness and localization in damaged materials. Clearly, it is considered that damage affects the shear behavior only[2]. Various expressions of $\mu(d)$ can be considered like [10.39] for the dilute scheme, [10.60] for the Mori–Tanaka scheme, and [12.79] for the PCW bound.

Considering the spherical symmetry of the problem, the displacement is sought in the radial form $\underline{\xi} = \xi(r)\underline{e}_r$ and spherical coordinates are adopted. Although the boundary condition on the external surface $r = b$ is written in terms of stress, the loading process is driven in terms of displacement. In other words, the radial displacement $\xi(b)$ is the quantity to which a *monotonic* evolution in time is prescribed. For instance, in the case $\dot{\xi}(b) > 0$, this makes it possible to investigate the softening regime in which the applied radial

2 In view of the developments in sections 10.4 and 10.5.2, this could be interpreted as the response of a microcracked material with randomly oriented closed frictionless microcracks, the damage variable d being the microcracks density parameter.

stress T on the external surface decreases even as the radial displacement increases.

The Navier equation $\mathrm{div}\,(\mathbb{C}(d) : \nabla \underline{\xi}) = 0$ for the damaged solid takes the form of an ordinary second-order differential equation coupling the two unknown functions $\xi(r)$ and $d(r)$:

$$(k + \frac{4}{3}\mu)\frac{\partial}{\partial r}\left(\frac{\partial \xi}{\partial r} + 2\frac{\xi}{r}\right) + \frac{4}{3}\frac{\partial \mu}{\partial d}\frac{\partial d}{\partial r}\left(\frac{\partial \xi}{\partial r} - \frac{\xi}{r}\right) = 0 \quad [13.27]$$

As in [13.9], the onset of damage is written here in terms of stress in the form:

$$\frac{1}{2}\boldsymbol{\sigma} : \frac{\partial}{\partial d}\mathbb{S}(d) : \boldsymbol{\sigma} = \mathcal{Y}_c \quad [13.28]$$

where $\mathbb{S}(d)$ is the compliance tensor. The assumption of constant bulk modulus yields

$$\frac{\partial}{\partial d}\mathbb{S}(d) = -\frac{1}{2\mu^2(d)}\frac{\partial \mu}{\partial d}\mathbb{K} \quad [13.29]$$

Accordingly, the damage criterion only involves the deviatoric stress $\boldsymbol{\sigma}_d = \mathbb{K} : \boldsymbol{\sigma}$ through the scalar stress invariant $\boldsymbol{\sigma}_d : \boldsymbol{\sigma}_d$:

$$\boldsymbol{\sigma}_d : \boldsymbol{\sigma}_d = \frac{2}{3}\left(\sigma_{rr} - \sigma_{\theta\theta}\right)^2 \quad [13.30]$$

In turn, the state equation relates the stress components to the displacement field:

$$\sigma_{rr} = (k - \frac{2}{3}\mu(d))\left(\frac{\partial \xi}{\partial r} + 2\frac{\xi}{r}\right) + 2\mu(d)\frac{\partial \xi}{\partial r} \quad [13.31]$$

$$\sigma_{\theta\theta} = (k - \frac{2}{3}\mu(d))\left(\frac{\partial \xi}{\partial r} + 2\frac{\xi}{r}\right) + 2\mu(d)\frac{\xi}{r} \qquad [13.32]$$

Eventually, the damage criterion takes the form of a second coupled differential equation:

$$-\frac{2}{3}\frac{\partial \mu}{\partial d}\left(\frac{\partial \xi}{\partial r} - \frac{\xi}{r}\right)^2 = \mathcal{Y}_c \qquad [13.33]$$

The initial damage yield: as long as this criterion is not reached, the response of the sphere remains linear elastic. The classical elastic solution reads:

$$\xi(r) = Ar + \frac{B}{r^2} \qquad [13.34]$$

and

$$\sigma_{rr} = 3(k - \frac{2}{3}\mu_o)A + 2\mu_o(A - 2\frac{B}{r^3}) \qquad [13.35]$$

and

$$\sigma_{\theta\theta} = \sigma_{rr} + 6\mu_o\frac{B}{r^3} = 3(k - \frac{2}{3}\mu_o)A + 2\mu_o(A + \frac{B}{r^3}) \qquad [13.36]$$

with

$$A = \frac{T}{3k}\frac{b^3}{b^3 - a_o^3}; \quad B = \frac{T}{4\mu_o}\frac{a_o^3 b^3}{b^3 - a_o^3} \qquad [13.37]$$

After damage onset, [13.34]–[13.36] remain valid in the elastic zone, A and B being determined by appropriate continuity conditions at the boundary between damaged and elastic zones. The damage criterion is reached at the cavity wall first. The onset of damage is characterized by $T = T_{el}$:

$$T_{el} = \frac{4\mu_o}{3}\left(\frac{a_o^3}{b^3} - 1\right)\zeta \quad \text{with} \quad \zeta = \epsilon\sqrt{-\frac{3\mathcal{Y}_c}{2\mu'_o}} \qquad [13.38]$$

where $\mu'_o = \partial\mu/\partial d_{|d=0}$, whereas $\epsilon = -1$ for $T > 0$ and $\epsilon = +1$ for $T < 0$. Again, the elastic response defined by [13.34] and [13.35]–[13.37] is valid as long as $|T| < |T_{el}|$. The displacement of the external sphere $r = b$ is given by:

$$\xi(b) = \frac{T(4b^3\mu_o + 3ka^3)}{12k\mu_o b^3 a^3} b \qquad [13.39]$$

Beyond the threshold $|T| = |T_{el}|$, damage is expected to affect an annular region $a_o < r < c$ for which an appropriate solution must be found.

13.3.2. *Case of damage model based on the dilute estimate*

We first consider an affine expression of the damaged shear modulus as suggested by the dilute estimate [10.39]:

$$\mu(d) = \mu_o(1 - Qd) \qquad [13.40]$$

in which we now denote by μ_o the shear modulus of the undamaged material[3].

Equation [13.33] then provides a first-order ordinary differential equation with respect to $\xi(r)$:

$$\frac{\partial\xi}{\partial r} - \frac{\xi}{r} = \epsilon\sqrt{\frac{3\mathcal{Y}_c}{2\mu_o Q}} = \zeta \qquad [13.41]$$

3 Note that the standard isotropic damage model corresponds to $Q = 1$.

The solution involves a constant u_o and reads:

$$\xi(r) = \zeta r \log r + u_o r \qquad (\epsilon = \pm 1) \qquad [13.42]$$

Introducing [13.42] into [13.27] yields the derivative of the damage variable $d(r)$:

$$\frac{\partial d}{\partial r} = \frac{3(3k + 4\mu)}{4\mu_o Q} \frac{1}{r} \qquad [13.43]$$

[13.43] reveals that $\partial d / \partial r$ is positive, which is clearly contrary to the physics of the problem in which the damage is expected to decrease from the cavity wall ($r = a_o$). Indeed, it is evident that it is not possible to derive a solution of the problem based on the dilute homogenization scheme. More precisely, it is not possible to meet the condition of monotonic increase (respectively, decrease) in traction (respectively, compression) of the loading parameter, namely $\xi(b)$. This is obviously a shortcoming of the damage model based on the dilute scheme in the present context where constant critical energy $\mathcal{Y}_c$ is considered.

13.3.3. *Complete solution in the case of the damage model based on PCW estimate*

Starting from the PCW estimate of the shear modulus, we more generally consider the following expression of $\mu(d)$:

$$\mu(d) = \mu_o \frac{1 + Qd}{1 + Q'd} \quad \text{with} \quad Q \leq 0 \leq Q' \qquad [13.44]$$

The PCW estimate [12.79] of the shear modulus actually corresponds to:

$$Q = -\frac{32}{225} \frac{7 - 5\nu_o}{2 - \nu_o}; \quad Q' = \frac{64}{225} \frac{4 - 5\nu_o}{2 - \nu_o} \qquad [13.45]$$

According to [13.44], the shear modulus becomes equal to 0 when the damage level reaches the critical value $-1/Q$. The physical evolutions of damage are therefore subjected to the restriction $d \leq -1/Q$.

Note that [13.44] also includes the predictions of the Mori–Tanaka scheme corresponding to

$$Q = 0; \quad Q' = \frac{32}{15}\frac{1-\nu_o}{2-\nu_o} \qquad [13.46]$$

for which the condition $\mu(d) > 0$ is always met.

13.3.3.1. *Closed-form expression of the mechanical field*

We now introduce [13.44] into [13.33]:

$$\frac{\partial \xi}{\partial r} - \frac{\xi}{r} = \zeta(1 + Q'd) \quad \text{with} \quad \zeta = \epsilon\sqrt{\frac{3\mathcal{Y}_c}{2\mu_o(Q' - Q)}} \qquad [13.47]$$

[13.47] is valid for $T \geq T_{el}$ (see equation [13.38]) in the damaged zone only. The solution $\xi(r)$ of [13.47] in the damaged zone can be placed in the form of an integral involving the damage function $d(r)$:

$$\xi(r) = v_o r + r\int_{a_o}^{r} \zeta(1 + Q'd(\rho))\frac{d\rho}{\rho} \qquad [13.48]$$

Introducing this expression into [13.27] yields an ordinary differential equation with respect to $d(r)$:

$$\frac{\partial d}{\partial r}\left(Q'(k + \frac{4}{3}\mu(d)) + \frac{4\mu_o(Q - Q')}{3(1 + Q'd)}\right) + 3(k + \frac{4}{3}\mu(d))(1 + Q'd)\frac{1}{r} = 0 \qquad [13.49]$$

which can be put in the form

$$\frac{\partial d}{\partial r} = -\frac{3}{r}\,(d + \frac{3k + 4\mu_o}{3kQ' + 4\mu_o Q}) \qquad [13.50]$$

As mentioned previously, the damage level d must remain below the limit value $-1/Q$ in order for the shear modulus to remain positive[4]. Combining this observation with the condition $\partial d/\partial r < 0$, the following compatibility requirement is necessary:

$$\frac{3k + 4\mu_o}{3kQ' + 4\mu_o Q} > \frac{1}{Q} \qquad [13.51]$$

Recalling that $Q \leq 0 \leq Q'$, [13.51] proves to be equivalent with the condition $Q(3kQ' + 4\mu_o Q) < 0$. This can only be achieved if

$$3kQ' + 4\mu_o Q > 0 \qquad [13.52]$$

For the PCW estimate, the (necessary) requirement $3kQ' + 4\mu_o Q > 0$ is fulfilled, if $\nu^s > 1/5$, for the considered class of shear moduli-based damage models. Otherwise, there is no physically meaningful response to the considered loading. These conditions are assumed to be fulfilled in the following.

The solution of [13.50] now reads:

$$d(r) = \frac{3k + 4\mu_o}{3kQ' + 4\mu_o Q}\left(\left(\frac{c}{r}\right)^3 - 1\right) \qquad [13.53]$$

4 In the case of the Mori–Tanaka scheme for which $Q = 0$, note that no restriction is prescribed to the damage level since the requirement $\mu(d) > 0$ is always satisfied.

For positive values of c, we note that the damage level decreases from the cavity wall, the minimum value of the shear modulus being reached at the cavity wall. In the above expression, c is defined by the condition $d(c) = 0$. It therefore denotes the radius that locates the boundary between the linear elastic zone ($r > c$) and the damaged zone ($r < c$). As already discussed, in order for the shear modulus to remain positive, the damage level d must remain below the limit value $-1/Q$. If $Q < 0$, the root $\tilde{a}$ of the equation $d(r) = -1/Q$ reads:

$$\tilde{a} = \mathcal{A}c \quad \text{with} \quad \mathcal{A} = \left(\frac{1 + 4\mu_o/3k}{1 - Q'/Q} \right)^{1/3} \qquad [13.54]$$

For later use, it is also convenient to introduce $c_{cr} = a_o/\mathcal{A}$, which represents the location of the boundary between damaged and elastic domains for which the initial cavity wall first meets the condition $\mu = 0$. Under the conditions [13.52], it appears that $\mathcal{A} < 1$, so that $c_{cr} > a_o$ and $\tilde{a} < c$.

If $\tilde{a}(c) < a_o$ (i.e. $c < c_{cr}$), the shear modulus satisfies the condition $\mu(d) > 0$ in the whole sphere ($a_o < r < c$). Otherwise, if $\tilde{a}(c) > a_o$ (i.e. $c > c_{cr}$), the region $a_o < r < \tilde{a}(c)$ is totally damaged insofar as it has lost its shear stiffness. The physics of the problem thus suggests that we remove this annular region, which results in writing the boundary condition $\sigma_{rr} = 0$ at $r = \tilde{a}$ instead of $r = a_o$. In the following, we denote by a the current cavity wall, on which the condition $\sigma_{rr} = 0$ is relevant. It is equal to a_o or $\tilde{a}(c)$, depending on the value of c:

$$a = \begin{cases} a_o & \text{if} \quad \tilde{a}(c) < a_o \\ \tilde{a}(c) & \text{if} \quad \tilde{a}(c) > a_o \end{cases} \qquad [13.55]$$

The expression of the displacement $\xi(r)$ in the damaged zone ($r < c$) is determined up to a constant u_o from a combination of [13.48] and [13.53]:

$$\xi(r) = u_o r - \frac{\zeta}{3(3kQ' + 4\mu_o Q)} \times (12\mu_o(Q' - Q) r \log r + Q'(3k + 4\mu_o)\frac{c^3}{r^2}) \qquad [13.56]$$

The constant u_o is determined from the condition $\sigma_{rr}(a) = 0$:

$$u_o = \frac{4\zeta\mu_o}{9k(3kQ' + 4\mu_o Q)}(9k(Q' - Q)\log a - (4\mu_o + 3k)Q\frac{c^3}{a^3}) \qquad [13.57]$$

The radial stress in the damaged zone is then determined from [13.31]:

$$\sigma_{rr} = \frac{12k(Q' - Q)\zeta\mu_o}{3kQ' + 4\mu_o Q}\log\frac{a}{r} + \frac{4\mu_o\zeta Q(3k + 4\mu_o)}{3(3kQ' + 4\mu_o Q)}\frac{c^3}{a^3}(\frac{a^3}{r^3} - 1) \qquad [13.58]$$

and $\sigma_{\theta\theta}$ is obtained from [13.32]

$$\sigma_{\theta\theta} = \sigma_{rr} + 2\mu(d)(\frac{\xi}{r} - \frac{\partial\xi}{\partial r}) \qquad [13.59]$$

the quantity $(\frac{\xi}{r} - \frac{\partial\xi}{\partial r})$ being given by [13.47].

We are left determining the relation between the loading defined by T and the location of the boundary of the damaged zone, defined by c. To do so, the following two cases must be considered:

– *Case 1:* $c < b$

In this case, the solid is divided into a damaged zone ($a < r < c$) and an elastic zone ($c < r < b$). In the damaged zone, the displacement is given by [13.56] and [13.57] and the radial stress is given by [13.58]. In the elastic zone, the displacement and the radial stress are determined from [13.34] and [13.35]. A and B are determined from the boundary condition $\sigma_{rr} = T$ and from the damage onset condition (see equation [13.47]):

$$\left\{\frac{\partial \xi}{\partial r} - \frac{\xi}{r}\right\}_{r=c} = \zeta \qquad [13.60]$$

The corresponding expression of the displacement of the external boundary $r = b$ reads:

$$\xi(b) = \frac{Tb}{3k} - c^3 \zeta \frac{3k + 4\mu_o}{9kb^2} \qquad [13.61]$$

In turn, the radial stress in the elastic zone is given by:

$$r > c: \quad \sigma_{rr} = T + \frac{4}{3}\mu_o \zeta c^3 \left(\frac{1}{r^3} - \frac{1}{b^3}\right) \qquad [13.62]$$

The link between T and c is obtained from the continuity of the radial stress in $r = c$. Combining [13.58] and [13.62] yields:

$$T = \frac{4}{3}\zeta\mu_o \left(\frac{3k(Q' - Q)}{3kQ' + 4\mu_o Q}(3\log\frac{a}{c} - 1) + \frac{c^3}{b^3} - \frac{c^3}{a^3}\frac{Q(3k + 4\mu_o)}{3kQ' + 4\mu_o Q}\right) \qquad [13.63]$$

with a given in [13.55]. Note that the continuity of the radial stress and of the deviatoric strain $\partial\xi/\partial r - \xi/r$ at $r = c$ also ensures the continuity of the displacement.

If $\tilde{a} > a_o$, that is $c > c_{cr}$, (together with $c < b$), the ratio a/c appearing in [13.63] is a constant. The expression of T reduces to an affine function of $\xi(b)$:

$$\tilde{a} > a_o : \quad \frac{T}{\mu_o} = \frac{4\zeta(Q' - Q)(3k + 4\mu_o)}{3kQ' + 4\mu_o Q} \log \mathcal{A} - 4\frac{\xi(b)}{b} \qquad [13.64]$$

– *Case 2*: $c > b$

According to the expression [13.53] of $d(r)$, the case $c > b$ corresponds to the situation in which the whole solid phase ($a < r < b$) is subjected to damage. This means that the radial stress is given by [13.58] on the whole interval $[a(c), b]$ (with $\sigma_{rr} = 0$ for $r < a(c)$). The link between the parameter c and T thus reads:

$$T = \frac{12k(Q' - Q)\zeta\mu_o}{3kQ' + 4\mu_o Q} \log \frac{a}{b} + \frac{4\mu_o\zeta Q(3k + 4\mu_o)}{3(3kQ' + 4\mu_o Q)} \frac{c^3}{a^3} (\frac{a^3}{b^3} - 1) [13.65]$$

The expression of the displacement is derived from [13.56] and [13.57].

$$\xi(b) = -\frac{b\zeta}{3kQ' + 4\mu_o Q} \times \left(4\mu_o(Q' - Q) \log \frac{b}{a} + \frac{4\mu_o + 3k}{9k} \frac{c^3}{a^3} (4\mu_o Q + 3kQ' \frac{a^3}{b^3})\right) \qquad [13.66]$$

13.3.3.2. *Maximum admissible load*

Assuming for a while that the maximum admissible loading $T_{\max}$ is reached in the regime $\tilde{a} < a_o$, the corresponding value

$c = c_{\max}$ is obtained from the condition[5] $\partial T/\partial c(c_{\max}) = 0$. From [13.63], we then obtain:

$$c_{\max} = a \left(\frac{3k(Q' - Q)}{-Q(3k + 4\mu_o) + (3kQ' + 4\mu_o Q)\frac{a^3}{b^3}} \right)^{1/3} \qquad [13.67]$$

In turn, under the conditions of [13.52], it is evident that $a_o < c_{\max} < c_{cr}$ as well as $c_{\max} < b$. $T_{\max}$ is then obtained from [13.63] with $a = a_o$ and $c = c_{\max}$. Physically speaking, this means that the peak value of the loading is reached:

– prior to total damage at the initial cavity wall;

– prior to damage reaching the external surface $r = b$.

13.3.3.3. *Macroscopic constitutive behavior: the T versus $\xi(b)$ curve*

The following two cases should be considered:

– *Case 1*: $c_{cr} < b$

Four different regimes will successively take place:

- Phase (1), $T < T_{el}$ (see [13.38]): This is the purely elastic regime. The macroscopic response is given by [13.39] (constant positive slope);

- Phase (2), $a_o < c < c_{cr}$: The domain $r \in [a_o, c]$ is subjected to partial damage, while the remainder $r \in [c, b]$ is still linear elastic; [13.63] (with $a = a_o$) and [13.61] are the parametric equations of this regime;

- Phase (3), $c_{cr} < c < b$: The domain $r \in [a_o, \tilde{a}(c)]$ is totally damaged. The macroscopic response is described by [13.64] (constant negative slope);

5 a can be regarded as a constant as long as the condition $\tilde{a} < a_o$, i.e. $c < c_{cr}$, is fulfilled.

- Phase (4), $b < c$: The whole structure is under partial or total damage; [13.65] and [13.66] are the parametric equations of this last phase that ends with failure.

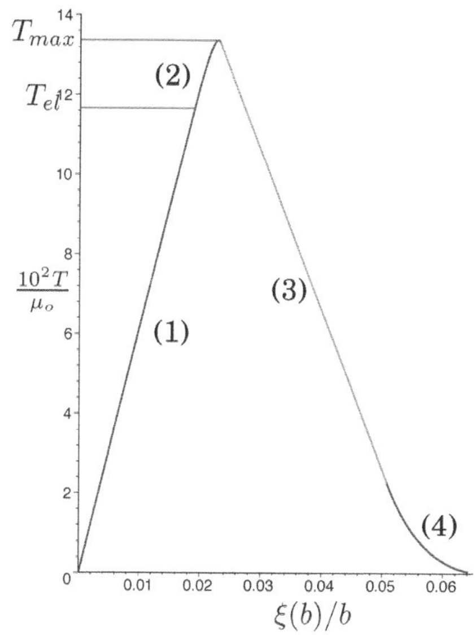

Figure 13.1. *Four-phase macroscopic response: T as a function of $\xi(b)/b$ (cas $c_{cr} < b$). $\zeta = -0.1$, $b/a = 2$, $k/\mu_o = 3$, $Q' = -Q = 1$*

– *Case 2*: $c_{cr} > b$

Again, four different regimes will successively take place:

- Phase (1), $T < T_{el}$ (see equation [13.38]): This is the purely elastic regime. The macroscopic response is given by [13.39] (constant positive slope);

- Phase (2), $a_o < c < b$: The domain $r \in [a_o, c]$ is subjected to partial damage, while the remainder $r \in [c, b]$ is still linear elastic. [13.63] (with $a = a_o$) and [13.61] are the parametric equations of this regime;

- Phase (3), $b < c < c_{cr}$: The whole domain $r \in [a, b]$ is subjected to partial damage. [13.65] and [13.66] with $a = a_o$ are the parametric equations of this phase. Eliminating the parameter c between these equations yields (constant negative slope):

$$T = \frac{12\mu_o k}{4b^3\mu_o Q + 3kQ'a^3} \times \left(Q(b^3 - a^3)\frac{\xi(b)}{b} - \zeta(Q' - Q)a^3 \log\frac{b}{a} \right) \qquad [13.68]$$

- Phase (4), $c > c_{cr}$ This last phase ends with failure. The parametric equations are [13.65] and [13.66] with $a = \tilde{a}(c)$.

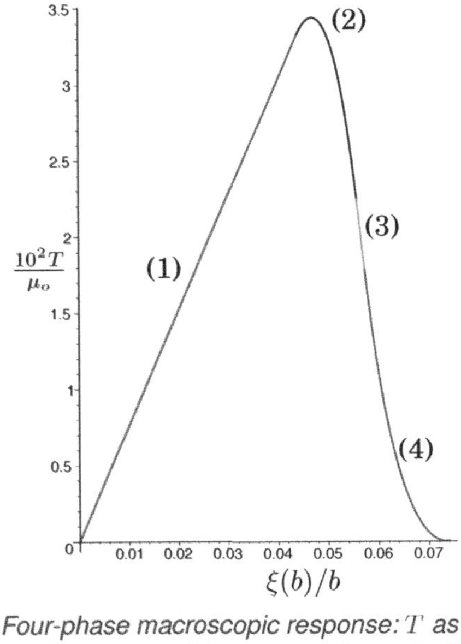

Figure 13.2. *Four-phase macroscopic response: T as a function of $\xi(b)$ (cas $c_{cr} > b$). $\zeta = -0.1$, $b/a_o = 1.1$, $k/\mu_o = 3$, $Q' = -Q = 1$*

13.3.3.4. *Solution in the unloading regime*

We consider a total unloading starting from a loading beyond the threshold T_{el}. The unloading phase is defined by the condition $\dot{\xi}(b) < 0$. The determination of the response of the structure is based on the assumption that the damage level remains constant during the unloading. In particular, the location $r = c$ of the boundary between damaged and undamaged zones is kept constant. We therefore deal with a heterogeneous structure in which the shear modulus $\mu(d)$ is defined by [13.44] together with [13.53] in the domain $r < c$, while $\mu = \mu_o$ in the domain $r > c$.

Accordingly, [13.34] and [13.35] still hold in the domain $r > c$. However, [13.47] is no more valid in the domain $r < c$. Instead, [13.27] should be considered, with $d(r)$ given by [13.53]. The general solution of the latter reads:

$$\xi(r) = Cr + D\left(12\mu_o(Q - Q')r\log r - (3k + 4\mu_o)Q'\frac{c^3}{r^2}\right) \qquad [13.69]$$

The four constants A, B and C, D, respectively, introduced in [13.34] and [13.69] are determined from:

– the boundary condition $\sigma_{rr}(b) = T$;

– the continuity of $\xi(r)$ and σ_{rr} at $r = c$;

– the boundary condition $\sigma_{rr}(a) = 0$.

These conditions produce a system of four linear equations with respect to the unknowns A, B and C, D. The solution reads:

$$A = \frac{Tb^3}{3k}\frac{3a^3k(3\log\frac{c}{a}+1)(Q-Q')-Qc^3(3k+4\mu_o)}{a^3\left(9b^3k(Q-Q')\log\frac{c}{a}+3b^3k(Q-Q')+c^3(3kQ'+4\mu_oQ)\right)-Qb^3c^3(3k+4\mu_o)} \qquad [13.70]$$

$$B = -\frac{Ta^3b^3c^3}{4\mu_o} \times \frac{3kQ'+4\mu_oQ}{a^3\left(9b^3k(Q-Q')\log\frac{c}{a}+3b^3k(Q-Q')+c^3(3kQ'+4\mu_oQ)\right)-Qb^3c^3(3k+4\mu_o)} \qquad [13.71]$$

$$C = -\frac{Tb^3}{3k} \times \frac{9a^3k(Q-Q')\log a + c^3(3k+4\mu_o)Q}{a^3\left(9b^3k(Q-Q')\log\frac{c}{a} + 3b^3k(Q-Q') + c^3(3kQ'+4\mu_oQ)\right) - Qb^3c^3(3k+4\mu_o)} \quad [13.72]$$

$$D = \frac{Ta^3b^3}{4\mu_o} \times \frac{1}{a^3\left(9b^3k(Q-Q')\log\frac{c}{a} + 3b^3k(Q-Q') + c^3(3kQ'+4\mu_oQ)\right) - Qb^3c^3(3k+4\mu_o)} \quad [13.73]$$

The expression of the displacement in the damaged zone ($r < c$) as a function of the current radial stress T reads:

$$\xi_{dam} = \frac{Tb^3}{12k\mu_o r^2} \times \frac{36a^3r^3k\mu_o(Q-Q')\log\frac{r}{a} - c^3(3k+4\mu_o)(4r^3\mu_oQ + 3a^3kQ')}{a^3\left(9b^3k(Q-Q')\log\frac{c}{a} + 3b^3k(Q-Q') + c^3(3kQ'+4\mu_oQ)\right) - Qb^3c^3(3k+4\mu_o)} \quad [13.74]$$

In turn, in the undamaged (homogeneous) zone ($r > c$), the classical expression [13.34] takes the form:

$$\xi_{el} = \frac{Tb^3}{3k}\,\frac{\left(3a^3k(3\log\frac{c}{a}+1)(Q-Q') - Qc^3(3k+4\mu_o)\right)r + \frac{3k}{4\mu_o}(3kQ'+4\mu_oQ)a^3c^3/r^2}{a^3\left(9b^3k(Q-Q')\log\frac{c}{a} + 3b^3k(Q-Q') + c^3(3kQ'+4\mu_oQ)\right) - Qb^3c^3(3k+4\mu_o)} \quad [13.75]$$

13.4. Stability of the solution to damage evolution in a hollow sphere

Hill's theory of stability for the solution of an evolution problem was initially formulated in the framework of elastoplasticity [HIL 58]. It can be readily extended to damage since the formulation only requires characterization of the incremental behavior by means of a tangent operator.

In this case, the latter takes two possible values depending on the damage rate $\dot{d}$:

$$\begin{aligned} &\dot{d} > 0 \Rightarrow \mathbb{C}_t = \mathbb{C}(d) - 2\frac{(\mathbb{C}'(d) : \boldsymbol{\varepsilon}) \otimes (\mathbb{C}'(d) : \boldsymbol{\varepsilon})}{\boldsymbol{\varepsilon} : \mathbb{C}''(d) : \boldsymbol{\varepsilon}} \\ &\dot{d} = 0 \Rightarrow \mathbb{C}_t = \mathbb{C}(d) \end{aligned} \quad [13.76]$$

where (see equation [13.26])

$$\mathbb{C}'(d) : \boldsymbol{\varepsilon} = 2\mu'(d)\mathbb{K} : \boldsymbol{\varepsilon} = 2\mu'(d)\boldsymbol{\varepsilon}_d$$

In the case of damage increase, the damage rate $\dot{d}$ and the strain rate $\mathbf{d} = \dot{\varepsilon}$ are related by:

$$\left(\frac{1}{2}\varepsilon : \mathbb{C}''(d) : \varepsilon\right)\dot{d} + \dot{\varepsilon} : \mathbb{C}'(d) : \boldsymbol{\varepsilon} = 0$$

so that the loading condition reads:

$$\dot{\varepsilon} : \mathbf{n} > 0 \quad \text{with} \quad \mathbf{n} = -\frac{\mathbb{C}'(d) : \boldsymbol{\varepsilon}}{\boldsymbol{\varepsilon} : \mathbb{C}''(d) : \boldsymbol{\varepsilon}} \quad [13.77]$$

In the present problem of a sphere subjected to traction on its external boundary, we have seen that

$$\boldsymbol{\varepsilon}_d = \epsilon_d \mathbf{N} \quad \text{with} \quad \mathbf{N} = \frac{1}{3}(2\underline{e}_r \otimes \underline{e}_r - \underline{e}_\theta \otimes \underline{e}_\theta - \underline{e}_\varphi \otimes \underline{e}_\varphi)$$

with

$$\epsilon_d = \frac{\partial \xi}{\partial r} - \frac{\xi}{r}$$

given by [13.47]. Traction being considered, we note that $\epsilon_d < 0$.

Let $\partial\Omega_\xi$ denote the part of the boundary of the structure Ω being considered on which the displacement is prescribed. Hill's theory of stability introduces the set $\mathcal{K}$ of virtual kinematically admissible velocity fields such that:

$$\underline{\hat{v}} \in \mathcal{K} \Leftrightarrow \begin{cases} \underline{\hat{v}}_{|\partial\Omega_\xi} = 0 \\ \underline{\hat{v}} \neq 0 \end{cases} \qquad [13.78]$$

Hill's sufficient condition for stability states that

$$(\forall \underline{\hat{v}} \in \mathcal{K}) \quad \mathcal{H}(\underline{\hat{v}}) = \int_\Omega \hat{\mathbf{d}} : \mathbb{C}_t : \hat{\mathbf{d}} \, dV > 0 \qquad [13.79]$$

in which $\hat{\mathbf{d}}$ is the virtual strain rate associated with $\underline{\hat{v}}$. In this case, we have:

$$\hat{\mathbf{d}} : \mathbb{C}_t : \hat{\mathbf{d}} = k(\text{div } \underline{\hat{v}})^2 + 2\mu(d) \; \hat{\mathbf{d}}_d : \hat{\mathbf{d}}_d - 6\frac{\mu'(d)^2}{\mu''(d)}(\mathbf{N} : \hat{\mathbf{d}}_{\mathbf{d}})^2$$

The Cauchy–Schwarz inequality further shows that:

$$(\mathbf{N} : \hat{\mathbf{d}}_d)^2 \leq (\mathbf{N} : \mathbf{N})(\hat{\mathbf{d}}_d : \hat{\mathbf{d}}_d) = \frac{2}{3} \hat{\mathbf{d}}_d : \hat{\mathbf{d}}_d$$

so that

$$\hat{\mathbf{d}} : \mathbb{C}_t : \hat{\mathbf{d}} \geq k(\text{div } \underline{\hat{v}})^2 + 2\left(\mu(d) - 2\frac{\mu'(d)^2}{\mu''(d)}\right) \hat{\mathbf{d}}_{\mathbf{d}} : \hat{\mathbf{d}}_{\mathbf{d}} \qquad [13.80]$$

13.4.1. *The MT damage model*

The reasoning presented hereafter is devoted to the particular situation [13.44]–[13.46]. In this case, it is evident that $\mu(d) = 2\mu'(d)^2/\mu''(d)$ so that

$$\hat{\mathbf{d}} : \mathbb{C}_t : \hat{\mathbf{d}} \geq k(\text{div } \underline{\hat{v}})^2 \qquad [13.81]$$

which holds in the loading condition ($\dot{\hat{d}} > 0$) as well as reversible condition ($\dot{\hat{d}} = 0$). We investigate whether it is possible to exhibit a counterexample of condition [13.79]. Owing to the positiveness of $\mathbb{C}_t$ stated by [13.81], this clearly implies that the condition $\hat{\mathbf{d}} : \mathbb{C}_t : \hat{\mathbf{d}} = 0$ must be satisfied in the whole structure. This amounts to the incompressibility condition:

$$\hat{\mathbf{d}} : \mathbb{C}_t : \hat{\mathbf{d}} = 0 \Rightarrow \text{div } \underline{\hat{v}} = 0$$

In the system of spherical coordinates, the following differential equation is obtained:

$$\frac{\partial \hat{v}_r}{\partial r} + 2\frac{\hat{v}_r}{r} = f \quad \text{with} \quad f = -\frac{1}{r\sin\theta}\frac{\partial \hat{v}_\varphi}{\partial \varphi} - \frac{1}{r}\frac{\partial \hat{v}_\theta}{\partial \theta} - \frac{\hat{v}_\theta \cot\theta}{r} \qquad [13.82]$$

As previously stated, the second requirement for $\hat{\mathbf{d}} : \mathbb{C}_t : \hat{\mathbf{d}} = 0$ is the loading condition $\dot{\hat{d}} > 0$ (otherwise, $\mathbb{C}_t$ takes the value $\mathbb{C}(d)$ that is definite positive); $\dot{\hat{d}}$ and $\hat{\mathbf{d}}$ are related by the damage criterion in rate form:

$$\left(\frac{1}{2}\varepsilon : \mathbb{C}''(d) : \varepsilon\right)\dot{\hat{d}} + \hat{\mathbf{d}} : \mathbb{C}'(d) : \boldsymbol{\varepsilon} = 0$$

that is:

$$\left(\mu''(d)\varepsilon_d : \varepsilon_d\right)\dot{\hat{d}} + 2\mu'(d)\hat{\mathbf{d}}_d : \boldsymbol{\varepsilon}_d = 0$$

so that (recalling $\epsilon_d < 0$):

$$\dot{\hat{d}} > 0 \Rightarrow \hat{\mathbf{d}}_d : \boldsymbol{\varepsilon}_d > 0 \Leftrightarrow \hat{\mathbf{d}}_d : \mathbf{N} < 0$$

This results in

$$3\hat{\mathbf{d}}_d : \mathbf{N} = 2\left(\frac{\partial \hat{v}_r}{\partial r} - \frac{\hat{v}_r}{r}\right) + f < 0 \qquad [13.83]$$

Combining [13.82] and [13.83] produces:

$$\frac{\partial \hat{v}_r}{\partial r} + 2\frac{\hat{v}_r}{r} < 2\left(\frac{\hat{v}_r}{r} - \frac{\partial \hat{v}_r}{\partial r}\right) \quad \Rightarrow \quad \frac{\partial \hat{v}_r}{\partial r} < 0 \qquad [13.84]$$

The boundary condition $\hat{v}_r(b) = 0$ implies that $\hat{v}_r(r, \theta, \varphi)$ is a positive decreasing function of r. Its maximum is therefore reached for $r = a$. In turn, the condition $\hat{v}_r(b) = 0$ together with $\operatorname{div} \underline{\hat{v}} = 0$ yields:

$$\int_{r=a} \hat{v}_r \, dS = 0 \qquad [13.85]$$

This can only be achieved if $\hat{v}_r(a, \theta, \varphi) = 0$, meaning that $\hat{v}_r(r, \theta, \varphi) = 0$ on the whole sphere. Placing this result into [13.82] further implies that $f = 0$. Eventually, we conclude that the loading condition [13.83] can never be satisfied by a velocity field meeting the incompressibility condition $\operatorname{div} \underline{\hat{v}} = 0$. In other words, it is not possible to exhibit a counterexample: This proves that Hill's sufficient condition of stability [13.79] is achieved for the MT-based damage model.

It is instructive to compare this conclusion to the 1D example in section 13.2. In the case of a MT damage model with $\mathcal{Y}_c = \text{Const.}$, we find that the tangent modulus $C_t(E_u, d_u) = 0$ (see equation [13.24]). In other words, the Hill's sufficient condition is not achieved.

13.4.2. *The general damage model* [13.44]

The damage model is now defined by [13.44] so that [13.80] becomes

$$\hat{\mathbf{d}} : \mathbb{C}_t : \hat{\mathbf{d}} \geq k(\operatorname{div} \underline{\hat{v}})^2 + 2\frac{\mu_o Q}{Q'} \hat{\mathbf{d}}_{\mathrm{d}} : \hat{\mathbf{d}}_{\mathrm{d}} \qquad [13.86]$$

in which we recall that $Q \leq 0$ and $Q' > 0$. We restrict to radial virtual velocity fields for which

$$\text{div } \underline{\hat{v}} = \frac{\hat{v}_r}{r} + 2\frac{\partial \hat{v}_r}{\partial r}; \quad \hat{\mathbf{d}}_{\mathrm{d}} : \hat{\mathbf{d}}_{\mathrm{d}} = \frac{2}{3}\left(\frac{\hat{v}_r}{r} - \frac{\partial \hat{v}_r}{\partial r}\right)^2$$

Under the loading condition ($\dot{\hat{d}} > 0$), [13.86] in fact reduces to

$$\hat{\mathbf{d}} : \mathbb{C}_t : \hat{\mathbf{d}} = k\left(\left(\frac{\hat{v}_r}{r} + 2\frac{\partial \hat{v}_r}{\partial r}\right)^2 - \eta\left(\frac{\hat{v}_r}{r} - \frac{\partial \hat{v}_r}{\partial r}\right)^2\right) \quad \text{with}$$

$$\eta = -\frac{4}{3}\frac{\mu_o Q}{kQ'} \tag{13.87}$$

In contrast, in the elastic region, we have $\mathbb{C}_t = \mathbb{C}(d = 0)$ and

$$\hat{\mathbf{d}} : \mathbb{C}_t : \hat{\mathbf{d}} = k\left(\left(\frac{\hat{v}_r}{r} + 2\frac{\partial \hat{v}_r}{\partial r}\right)^2 + \eta_o\left(\frac{\hat{v}_r}{r} - \frac{\partial \hat{v}_r}{\partial r}\right)^2\right) \quad \text{with}$$

$$\eta_o = \frac{4\mu_o}{3k} \tag{13.88}$$

Following [13.79], a sufficient condition of stability reads:

$$(\forall\, \hat{v}_r \neq 0,\ \hat{v}_r(b) = 0) \quad \mathcal{H}(\hat{v}(r)\underline{e}_r) > 0 \tag{13.89}$$

that is

$$\begin{aligned}\tilde{\mathcal{H}}(\hat{v}) = &\int_a^c \left(\left(\frac{\hat{v}}{r} + 2\frac{\partial \hat{v}}{\partial r}\right)^2 - \eta\left(\frac{\hat{v}}{r} - \frac{\partial \hat{v}}{\partial r}\right)^2\right) dr + \\ &\int_c^b \left(\left(\frac{\hat{v}}{r} + 2\frac{\partial \hat{v}}{\partial r}\right)^2 + \eta_o\left(\frac{\hat{v}}{r} - \frac{\partial \hat{v}}{\partial r}\right)^2\right) dr > 0\end{aligned} \tag{13.90}$$

If this sufficient condition of stability is not fulfilled, there exists a velocity field $\hat{v}(r)$ for which $\tilde{\mathcal{H}}(\hat{v}) \leq 0$. Since $\tilde{\mathcal{H}}(\lambda\hat{v}) = \lambda^2\tilde{\mathcal{H}}(\hat{v})$, it is not restrictive to assume that $\hat{v}(a) = 1$. Therefore let $\mathcal{K}_r = \{\hat{v}(r) \neq 0,\, \hat{v}(a) = 1,\, \hat{v}(b) = 0\}$ and consider the variational problem:

$$H = \inf_{v(r)\in\mathcal{K}_r} \tilde{\mathcal{H}}(\hat{v}) \qquad [13.91]$$

It is evident that the sufficient condition of stability is equivalent to $H > 0$. Let $v^*(r)$ denote the velocity field for which $\tilde{\mathcal{H}}(v^*(r)) = H$. Standard calculus of variations shows that the sought field $v^*(r)$ is the solution to the same second-order differential equation in both intervals[6] $[a, c]$ and $[c, b]$, namely:

$$\frac{d^2u}{dr^2} + \frac{2}{r}\frac{du}{dr} - 2\frac{u}{r^2} = 0 \qquad [13.92]$$

whose solutions are of the form $u(r) = C_1 r + C_2/r^2$. Note, however, that the constants C_i depend on the considered interval. Since $v^*(a) = 1$ and $v^*(b) = 0$, the optimization amounts to determine the value of $v^*(c)$ that minimizes $\tilde{\mathcal{H}}(\hat{v})$. More precisely, we consider the velocity fields defined piecewise as follows:

$$\begin{aligned} r \in [a, c] &: \hat{v}(r) = \beta\left(r - \frac{a^3}{r^2}\right) + \frac{a^2}{r^2} \\ r \in [c, b] &: \hat{v}(r) = \alpha\left(r - \frac{b^3}{r^2}\right) \end{aligned} \qquad [13.93]$$

with the continuity condition at $r = c$ that yields:

$$\beta = \frac{\alpha(b^3 - c^3) + a^2}{a^3 - c^3} \qquad [13.94]$$

6 We recall that $r = c$ is the boundary between the linear elastic and damaged zones.

The optimization can be performed conveniently with respect to α. The optimal value is

$$\alpha^* = -\frac{1}{a}\frac{\varphi}{\eta_o + 1} \qquad [13.95]$$

where $\varphi = a^3/b^3$ represents the cavity volume fraction (i.e. the porosity). For this value α^*, the minimum $\tilde{\mathcal{H}}(v^*(r)) = H$ is reached. Stability is therefore granted below the critical threshold $\eta < \eta_c$ that is the solution to the equation $H = 0$ and given by

$$\eta_c = \frac{\varphi\rho^3 + \eta_o}{\rho^3(\varphi + \eta_o + 1) - 1} \qquad [13.96]$$

In [13.96], $\rho = c/a$ characterizes the extension of the damaged zone. For example, at the onset of damage, we have $\rho = 1$ that corresponds to $\eta_c = 1$. In turn, when damage reaches the external boundary, we have $\varphi\rho^3 = 1$ and [13.96] yields $\eta_c = \varphi$.

Conversely, for a given value of η, the extension of the damaged zone below which stability is granted is characterized by $\rho < \rho_c$:

$$\rho_c^3 = \frac{\eta + \eta_o}{\eta(\varphi + \eta_o + 1) - \varphi} \qquad [13.97]$$

which is a decreasing function of η. The critical value $\rho_c = 1$ is retrieved for $\eta = 1$, meaning that stability is not granted at the onset of damage if $\eta \geq 1$; [13.96] also reveals that η_c is an increasing function of η_o, meaning that a lower bound of η_c is obtained at the limit $\eta_o \to 0$ and takes the form $\eta_c^{\ell b} = 1/\rho^3$. Figure 13.3 presents the variation of η_c and this lower bound ($\varphi = 0.3$, $\eta_o = 1$).

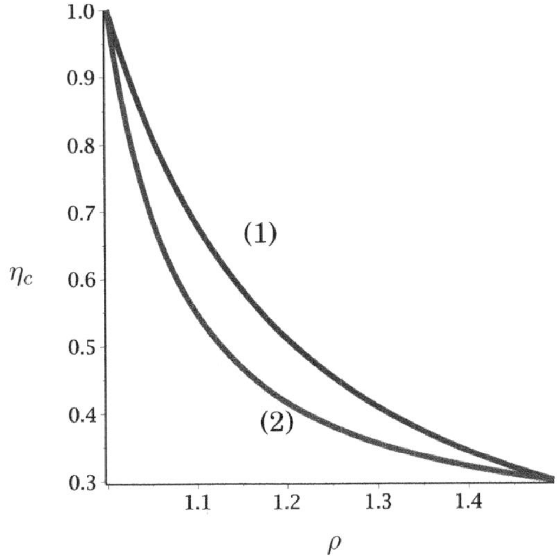

Figure 13.3. *Evolution of the critical threshold* η_c. $\varphi = 0.3$, *Curve (1):*$\eta_o = 1$. *Curve (2):* $\eta_o = 0$

Interestingly, the optimal field v^* does not *a priori* vanish in the elastic domain, as indicated by the value of α^* (see [13.95]). However, in the limit case $\eta_o \rightarrow 0$, we note that $\alpha^* \rightarrow 0$, meaning that the optimal field vanishes out of the damaged zone.

Bibliography

[AND 86] ANDRIEUX S., BAMBERGER Y., MARIGO J., "Un modèle de matériau microfissuré pour les roches et les bétons", *Journal de Mécanique Théorique et Appliquée*, vol. 5, no. 3, pp. 471–513, 1986.

[AZH 00] AZHDARI A., OBATA M., NEMAT-NASSER S., "Alternative solution methods for crack problems in plane anisotropic elasticity, with examples", *International Journal of Solids and Structures*, vol. 37, pp. 6433–6478, 2000.

[BEN 87] BENVENISTE Y., "A new approach to the application of mori-tanaka's theory in composite materials", *Mechanics of Materials*, vol. 6, pp. 147–157, 1987.

[BUD 76] BUDIANSKY B., O'CONNELL R., "Elastic moduli of a cracked solid", *International Journal of Solids Structure*, vol. 12, pp. 81–97, 1976.

[BUI 78] BUI H.D., *Mécanique de la rupture fragile*, Masson, Paris, 1978.

[BUI 06] BUI H.D., *Fracture mechanics: inverse problems and solutions*, Springer, Dordrecht, the Netherlands, 2006.

[DED 73] DEDERICHS P.H., ZELLER R., "Variational treatment of the elastic constants of disordered materials", *Zeitschrift für Physik*, vol. 259, pp. 103–116, 1973.

[DES 94] DESOYER T., CORMERY F., "On uniqueness and localization in elastic-damage materials", *International Journal of Solids Structure*, vol. 31, pp. 733–744, 1994.

[DEU 02] DEUDÉ V., DORMIEUX L., KONDO D. *et al.*, "Micromechanical approach to nonlinear poroelasticity: application to cracked rocks", *Journal of Engineering Mechanics*, vol. 128, pp. 848–855, 2002.

[DOR 05] DORMIEUX L. KONDO D., "Poroelasticity and damage theory for saturated cracked media", in DORMIEUX L., ULM F.J., (eds), *Applied Micromechanics of Porous Materials*, Springer, New York, 2005.

[DOR 06] DORMIEUX L., KONDO D., ULM F.-J., *Microporomechanics*, Wiley, 2006.

[DOR 09] DORMIEUX L., KONDO D., "Stress-based estimates and bounds of effective elastic properties: the case of cracked media with unilateral effects", *Computational Materials Science*, vol. 46, pp. 173–179, 2009.

[ESH 57] ESHELBY J.D., "The determination of the elastic field of an ellipsoidal inclusion, and related problems", *Proceedings of the Royal Soceity of London*, vol. A 241, pp. 376–396, 1957.

[GRI 21] GRIFFITH A.A., "The phenomenon of rupture and flow in solids", *Philosophical Transactions of the Royal Society of London*, vol. 221, pp. 163–198, 1921.

[HAS 83] HASHIN Z., "Analysis of composite materials: a survey", *Journal of Applied Mechanics*, vol. 50, pp. 481–505, 1983.

[HAS 63] HASHIN Z., SHTRIKMAN S., "A variational approach to the theory of the elastic behavior of multiphase materials", *Journal of Mechanics and Physics of Solids*, vol. 11, pp. 127–140, 1963.

[HIL 58] HILL R., "A general theory of uniqueness and stability in elasto-plastic solids", *International Journal of Mechanics and Physics of Solids*, vol. 6, pp. 236–249, 1958.

[HIL 67] HILL R., "The essential structure of constitutive laws for metal composites and polycrystals", *Journal of Mechanics and Physics of Solids*, vol. 15, pp. 79–95, 1967.

[HOR 83] HORII H., NEMAT-NASSER S., "Overall modulii of solids with microcracks: load-induced anisotropy", *Journal of the Mechanics and Physics of Solids*, vol. 31, pp. 155–171, 1983.

[KAC 93] KACHANOV M., *Elastic Solids with Many Cracks Related Problems*, Academic Press, San Diego, 1993.

[KAC 03] KACHANOV M., SHAFIRO B., TSUKROV I., *Handbook of Elasticity Solutions*, Kluwer Academic Publishers, 2003.

[KEL 67] KELLOG O.D., *Foundations of Potential Theory*, Springer-Verlag, Berlin, 1967.

[KRA 96] KRAJCINOVIC D., *Damage Mechanics*, North-Holland, Amsterdam, 1996.

[LAW 85] LAWS N., "A note on penny-shaped cracks in transversely isotropic materials", *Mechanics of Materials*, vol. 4, pp. 209–212, 1985.

[LEB 03] LEBLOND J.-B., *Mécanique de la rupture fragile et ductile*, Hermès-Lavoisier, Paris, 2003.

[LEK 63] LEKHNITSKII S.G., *Theory of Elasticity of an Anisotropic Body*, MIR, Mosow, 1963.

[LIE 68] LIEBOWITZ H. (ed.), *Fracture, an Advanced Treatise. Mathematical Fundamentals*, vol. 2, Academic Press, 1968.

[MAN 72] MANDEL J., "Plasticité classique et viscoplasticité", *CISM Lecture Notes no. 97*, Springer, 1972.

[MOR 73] MORI T., TANAKA K., "Average stress in matrix and average elastic energy of materials with misfitting inclusions", *Acta Metallurgica*, vol. 21, p. 571, 1973.

[MUR 87] MURA T., *Micromechanics of Defects in Solids*, 2nd ed., Martinus Nijhoff Publishers, 1987.

[MUS 53] MUSKHELISHVILI N.Y., *Some Basic Problems of Mathematical Theory of Elasticity*, Noordhoff, Groningen, 1953.

[NEM 93] NEMAT-NASSER S., HORII M., *Micromechanics: Overall Properties of Heterogeneous Materials*, North-Holland, Amsterdam, 1993.

[PEN 02] PENSÉE V., KONDO D., DORMIEUX L., "Micromechanical Analysis of Anisotropic Damage in Brittle Materials", *Journal of Engineering Mechanics*, vol. 128, pp. 889–897, 2002.

[PON 95] PONTE-CASTANEDA P., WILLIS J.R., "The effect of spatial distribution on the effective behavior of composite materials and cracked media", *Journal of Mechanics and Physics of Solids*, vol. 43, no. 12, pp. 1919–1951, 1995.

[RIC 78] RICE J.R., "Thermodynamics of the quasi-static growth of griffith cracks", *Journal of the Mechanics and Physics of Solids*, vol. 26, pp. 61–78, 1978.

[SEG 50] SEGEDIN C.M., "Note on penny-shaped crack under shear", *Proceedings of the Cambridge Philosophical Society*, vol. 47, p. 396, 1950.

[SIH 65] SIH G.C., PARIS P.C., IRWIN G.R., "On cracks in rectilinearly anisotropic bodies", *International Journal of Fracture Mechanics*, vol. 1, pp. 189–203, 1965.

[SNE 69] SNEDDON I.N., LOWENGRUB M., *Crack Problems in the Classical Theory of Elasticity*, Wiley, 1969.

[WIL 97] TALBOT D.R.S., WILLIS J.R., "Bounds of third order for the overall response of non linear composites" *Journal of Mechanics and Physics of Solids*, vol. 45, pp. 87–111, 1997.

[WAL 81] WALPOLE L.J., "Elastic behavior of composite materials: theoretical foundations", *Advances in Applied Mechanics*, vol. 21, pp. 169–242, 1981.

[WIL 77] WILLIS J.R., "Bounds and self-consistent estimates for the overall properties of anisotropic composites", *Journal of Mechanics and Physics of Solids*, vol. 25, pp. 185–202, 1977.

[WIL 80] WILLIS J.R., "A polarization approach to the scattering of elastic waves-II. Multiple scattering from inclusions", *Journal of Mechanics and Physics of Solids*, vol. 28, pp. 307–327, 1980.

[WIL 81] WILLIS J.R., "Variational and related methods for the overall properties of composites", *Advances in Applied Mechanics*, vol. 21, pp. 1–78, 1981.

[ZAO 02] ZAOUI A., "Continuum micromechanics: survey", *Journal of Engineering Mechanics*, vol. 128, pp. 808–816, 2002.

Index

A, C, D

E, F

G, H, I

M, O, P

S, T, U

Other titles from

iSTE

in

Mechanical Engineering and Solid Mechanics

2016

BOYARD Nicolas
Heat Transfer in Polymer Composite Materials

DE SAXCÉ Géry
Galilean Mechanics and Thermodynamics of Continua

2015

KARLIČIĆ Danilo, MURMU Tony, ADHIKARI Sondipon, MCCARTHY Michael
Non-local Structural Mechanics

SAB Karam, LEBÉE Arthur
Homogenization of Heterogeneous Thin and Thick Plates

2014

ATANACKOVIC M. Teodor, PILIPOVIC Stevan, STANKOVIC Bogoljub, ZORICA Dusan
Fractional Calculus with Applications in Mechanics: Vibrations and Diffusion Processes

ATANACKOVIC M. Teodor, PILIPOVIC Stevan, STANKOVIC Bogoljub, ZORICA Dusan
Fractional Calculus with Applications in Mechanics: Wave Propagation, Impact and Variational Principles

CIBLAC Thierry, MOREL Jean-Claude
Sustainable Masonry: Stability and Behavior of Structures

ILANKO Sinniah, MONTERRUBIO Luis E., MOCHIDA Yusuke
The Rayleigh–Ritz Method for Structural Analysis

LALANNE Christian
Mechanical Vibration and Shock Analysis – 5-volume series – 3rd edition
Sinusoidal Vibration – volume 1
Mechanical Shock – volume 2
Random Vibration – volume 3
Fatigue Damage – volume 4
Specification Development – volume 5

LEMAIRE Maurice
Uncertainty and Mechanics

2013

ADHIKARI Sondipon
Structural Dynamic Analysis with Generalized Damping Models: Identification

ADHIKARI Sondipon
Structural Dynamic Analysis with Generalized Damping Models: Analysis

BAILLY Patrice
Materials and Structures under Shock and Impact

BASTIEN Jérôme, BERNARDIN Frédéric, LAMARQUE Claude-Henri
Non-smooth Deterministic or Stochastic Discrete Dynamical Systems: Applications to Models with Friction or Impact

EL HAMI Abdelkhalak, BOUCHAIB Radi
Uncertainty and Optimization in Structural Mechanics

KIRILLOV Oleg N., PELINOVSKY Dmitry E.
Nonlinear Physical Systems: Spectral Analysis, Stability and Bifurcations

LUONGO Angelo, ZULLI Daniele
Mathematical Models of Beams and Cables

SALENÇON Jean
Yield Design

2012

DAVIM J. Paulo
Mechanical Engineering Education

DUPEUX Michel, BRACCINI Muriel
Mechanics of Solid Interfaces

ELISHAKOFF Isaac *et al.*
Carbon Nanotubes and Nanosensors: Vibration, Buckling and Ballistic Impact

GRÉDIAC Michel, HILD François
Full-Field Measurements and Identification in Solid Mechanics

GROUS Ammar
Fracture Mechanics – 3-volume series
Analysis of Reliability and Quality Control – volume 1
Applied Reliability – volume 2
Applied Quality Control – volume 3

RECHO Naman
Fracture Mechanics and Crack Growth

2011

KRYSINSKI Tomasz, MALBURET François
Mechanical Instability

SOUSTELLE Michel
An Introduction to Chemical Kinetics

2010

BREITKOPF Piotr, FILOMENO COELHO Rajan
Multidisciplinary Design Optimization in Computational Mechanics

DAVIM J. Paulo
Biotribolgy

PAULTRE Patrick
Dynamics of Structures

SOUSTELLE Michel
Handbook of Heterogenous Kinetics

2009

BERLIOZ Alain, TROMPETTE Philippe
Solid Mechanics using the Finite Element Method

LEMAIRE Maurice
Structural Reliability

2007

GIRARD Alain, ROY Nicolas
Structural Dynamics in Industry

GUINEBRETIÈRE René
X-ray Diffraction by Polycrystalline Materials

KRYSINSKI Tomasz, MALBURET François
Mechanical Vibrations

KUNDU Tribikram
Advanced Ultrasonic Methods for Material and Structure Inspection

SIH George C. *et al.*
Particle and Continuum Aspects of Mesomechanics

Printed and bound by CPI Group (UK) Ltd, Croydon, CR0 4YY
13/09/2026
14961020-0006